NATO ASI Series

Advanced Science Institutes Series

A series presenting the results of activities sponsored by the NATO Science Committee, which aims at the dissemination of advanced scientific and technological knowledge, with a view to strengthening links between scientific communities.

The Series is published by an international board of publishers in conjunction with the NATO Scientific Affairs Division

A	Life Sciences	Plenum Publishing Corporation
B	Physics	London and New York
C	Mathematical and Physical Sciences	Kluwer Academic Publishers
D	Behavioural and Social Sciences	Dordrecht, Boston and London
E	Applied Sciences	
F	Computer and Systems Sciences	Springer-Verlag
G	Ecological Sciences	Berlin Heidelberg New York
H	Cell Biology	London Paris Tokyo Hong Kong
I	Global Environmental Change	Barcelona Budapest

PARTNERSHIP SUB-SERIES

1. Disarmament Technologies	Kluwer Academic Publishers
2. Environment	Springer-Verlag
3. High Technology	Kluwer Academic Publishers
4. Science and Technology Policy	Kluwer Academic Publishers
5. Computer Networking	Kluwer Academic Publishers

The Partnership Sub-Series incorporates activities undertaken in collaboration with NATO's Cooperation Partners, the countries of the CIS and Central and Eastern Europe, in Priority Areas of concern to those countries.

NATO-PCO DATABASE

The electronic index to the NATO ASI Series provides full bibliographical references (with keywords and/or abstracts) to about 50 000 contributions from international scientists published in all sections of the NATO ASI Series. Access to the NATO-PCO DATABASE compiled by the NATO Publication Coordination Office is possible in two ways:

- via online FILE 128 (NATO-PCO DATABASE) hosted by ESRIN,
  Via Galileo Galilei, I-00044 Frascati, Italy.

- via CD-ROM "NATO Science & Technology Disk" with user-friendly retrieval software in English, French and German (© WTV GmbH and DATAWARE Technologies Inc. 1992).

The CD-ROM can be ordered through any member of the Board of Publishers or through NATO-PCO, Overijse, Belgium.

Series F: Computer and Systems Sciences, Vol. 117

The ASI Series F Subseries on
ADVANCED EDUCATIONAL TECHNOLOGY

This book contains the proceedings of a NATO Advanced Research
Workshop held within the activities of the NATO Special Programme on
Advanced Educational Technology, running from 1988 to 1993 under the
auspices of the NATO Science Committee. The books published so far in
the Special Programme are listed briefly, as well as in detail together with
other volumes in NATO ASI Series F, at the end of this volume.

Springer-Verlag Berlin Heidelberg GmbH

Intelligent Learning Environments: The Case of Geometry

Edited by

Jean-Marie Laborde

Laboratoire de Structures Discrètes et de Didactique
Institut d'Informatique et de Mathématiques Appliquées de Grenoble
Université Joseph Fourier
Centre Nationale de la Recherche Scientifique
BP 53, F-38041 Grenoble cedex 9, France

Springer
Published in cooperation with NATO Scientific Affairs Division

Proceedings originating from the NATO Advanced Research Workshop on Intelligent Learning Environments: the Case of Geometry, held in Grenoble, France, November 13–16, 1989

Cataloging-in-Publication Data applied for

Die Deutsche Bibliothek - CIP-Einheitsaufnahme

Intelligent learning environments : the case of geometry ; [proceedings of the NATO Advanced Research Workshop on Intelligent Learning Environments: The Case of Geometry, held in Grenoble, France, November 13 - 16, 1989] / ed. by Jean-Marie Laborde. Publ. in cooperation with NATO Scientific Affairs Division. - Berlin ; Heidelberg ; New York ; Barcelona ; Budapest ; Hong Kong ; London ; Milan ; Paris ; Santa Clara ; Singapore ; Tokyo : Springer, 1996
  (NATO ASI series : Ser. F, Computer and systems sciences ; Vol. 117)
ISBN 978-3-642-64608-9   ISBN 978-3-642-60927-5 (eBook)
DOI 10.1007/978-3-642-60927-5

NE: Laborde, Jean-Marie [Hrsg.]; Advanced Research Workshop on Intelligent Learning Environments: The Case of Geometry <1989, Grenoble>; NATO: NATO ASI series / F

CR Subject Classification (1991): K.3, I.2, J.2

© Springer-Verlag Berlin Heidelberg 1996

Softcover reprint of the hardcover 1st edition 1996

Typesetting: Camera-ready by editor
Printed on acid-free paper
SPIN: 10084412   45/3142 – 5 4 3 2 1 0

Preface

This book is a thoroughly revised result, updated to mid-1995, of the NATO Advanced Research Workshop on "Intelligent Learning Environments: the case of geometry", held in Grenoble, France, November 13–16, 1989.

The main aim of the workshop was to foster exchanges among researchers who were concerned with the design of intelligent learning environments for geometry. The problem of student modelling was chosen as a central theme of the workshop, insofar as geometry cannot be reduced to procedural knowledge and because the significance of its complexity makes it of interest for intelligent tutoring system (ITS) development. The workshop centred around the following themes: modelling the knowledge domain, modelling student knowledge, designing "didactic interaction", and learner control.

This book contains revised versions of the papers presented at the workshop. All of the chapters that follow have been written by participants at the workshop. Each formed the basis for a scheduled presentation and discussion. Many are suggestive of research directions that will be carried out in the future. There are four main issues running through the papers presented in this book:

- knowledge about geometry is not knowledge about the real world, and materialization of geometrical objects implies a reification of geometry which is amplified in the case of its implementation in a computer, since objects can be manipulated directly and relations are the results of actions (Laborde, Schumann). This aspect is well exemplified by research projects focusing on the design of geometric microworlds (Guin, Laborde).

- the learning of geometry involves conceptualization at a number of different levels, such as those of geometrical "objects" and their relationships, conceptual constructs and general properties, high order thinking and mathematical proof (Chouraqui & Inghilterra, Hoyle, Gras & Giorgiutti). Thus modelling of student knowledge must be broad enough to capture student views about proving, generalizing, and so on (Yerulshalmy), whereas it classically focuses on procedural and declarative knowledge.

- the design of the interface as well as the management of didactical interaction must cope with the multiple register of representation required by geometry: drawing, figure description in natural language, numerical characterization, and so on (Straesser). The question of the design of didactical interaction has been considered taking into account similarities and differences between CAI and ICAI (Schwartz).

• domains of geometry other than plane geometry were considered during the workshop (Thibault & Labarre, Vivet), also the above problems were discussed in relation to projects related to other domains of knowledge (Devi et al., Dreyfus & Schwartz, Falcone).

This book did not appear immediately after the workshop was held in Grenoble. There are many reasons for this, which the reader may easily understand, for at this end of the millennium it is a common situation, unfortunately, to run out of time. One may wonder whether the content of the book reflects the current state of the art. My deep feeling is that some papers are global enough not to be really time-stamped; moreover the delay in publication has had at least one advantage, namely that the contributors have had the opportunity to update their papers, which finally means this book has still been published at the right time.

Acknowledgements

Organizing a workshop or editing a book is never a task one can do in isolation. So I would like to express my gratitude to several institutions and persons that helped me in making these events possible. First to the members of the Scientific Committee, C. Hoyles (Institute of Education, London) and J. Schwartz (MIT, Cambridge USA), and all participants. Their effort and enthusiasm before and during the meeting was fundamental for the success of the workshop. Second to the Institut d'Informatique et de Mathématiques Appliquées de Grenoble, the Laboratoire de Structures Discrètes et de Didactique, and finally for their commitment in the management of the workshop, all the members of the project CABRI-géomètre. Special thanks go to Yves Carbonneaux, working on the CABRI-graph project, who carefully edited and polished the whole manuscript. Without his contribution the present volume would certainly never have appeared. I would also like to thank the editors at Springer-Verlag for their patience and tolerance, especially J. Andrew Ross, who made a further round of stylistic and linguistic improvements to the book. There is one last person, who supported the project continuously from the very beginning of the ARW to the release of this volume, and who promoted the ideas needed to fulfill the entire project: on behalf of all the people involved in the ARW, I would like to thank you again, Nicolas Balacheff.

Grenoble   Jean-Marie Laborde
July 1995

Table of Contents

A Model of Case-Based Reasoning for Solving Problems of Geometry in a Tutoring System

Eugène Chouraqui and Carlo Inghilterra

Département de Recherche en Informatique Automatique Mecatronique
I.U.S.P.I.M. avenue Escadrille Normandie Niemen
F-13397 Marseille Cedex 20, France
E-mail: DIAM_EC@vmesa11.u-3mrs.fr, DIAM_CI@vmesa11.u-3mrs.fr

Abstract. One of the most important requirements of an intelligent tutoring systems for solving elementary problems in geometry is its ability to guide and follow a human learner. To achieve this task we assume that the teaching aid should operate in a way similar to the learner's activity. Human learning is based on stepwise cumulative experiments. Therefore, we introduce a model of case-based reasoning – a special analogical reasoning paradigm – that keeps a trace of past experience, so that it can use it for solving new problems analogous to those already memorized. After defining criteria for assessing the analogy of two problems we describe the evolution of Long Term Memory in the tutoring system. This method endows the system with an "apprentice appearance" facilitating its adaptation to human learners.

Keywords. Problem solving, Analogical reasoning, Case-based reasoning, Similarity ratio, Signature, Geometry, Frames, Machine learning, Tutoring systems

1. Introduction

Pedagogical experience has brought to evidence that many pupils face great difficulties in solving problems despite their knowledge of the geometric course (i.e., declarative knowledge). Still, they may progressively acquire a certain degree of competence in solving problems if they are exposed to many training exercises. Therefore it is legitimate to suppose that past resolutions imprint a trace on memory that may be activated to help solving new problems [PIN.91].

The tutoring system aimed at teaching how to solve elementary geometric problems comprises a Knowledge Base represented as frames and a Base of Facts containing problems to solve [CHO.87]. The Resolution Module assists learners in their own reasoning while monitoring their activities precisely. Therefore, one of the main functions of this tutoring system is to guide and follow the human learner while adapting itself to the learner's own processes in the task of solving problems.

2. A Cognitive Analysis of Problem Solving

Most tutoring systems based on expert system methods/technology produce a combinatorial trace of resolutions; this trace is too syntax-oriented for a learner who is in need for a reasonably argumented assistance [BLE.86]. These systems cannot learn as they retain almost none of their past experience. A difficult problem is solved using the same techniques as a beginner's exercise. This goes against one of the fundamental principles in human learning, which is the need of cumulating varied and progressive experience throughout one's training [CAU.85]. A person solving a problem with great competence in its domain will invoke experience that has already been assimilated and he/she considers to be related with the problem. Such a person will therefore be guided in doing choices relevant to the present context, trying plausible procedures and adequate heuristics that will limitate combinatorial explosion. In a similar way, a good teacher is one who is able to make his/her mental process explicit while solving problems. The teacher unwinds a sort of "Ariadne's thread" showing why and how he/she proceeded to the solution.

A good tutoring system, therefore, should be able to simulate at best cognitive activity involved in problem solving in order to follow the human learner's own resolution process, with the aim of better following, guiding and controlling the learner.

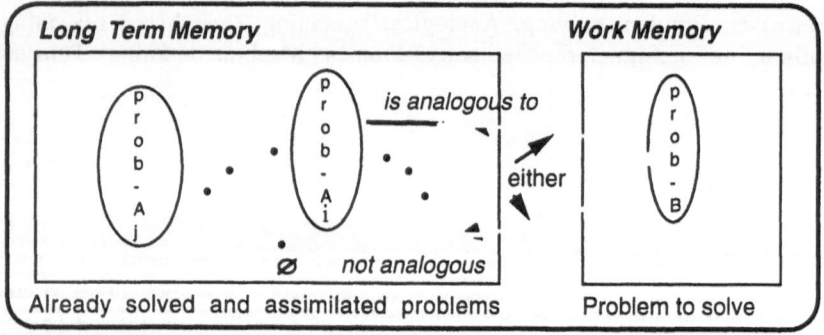

Fig. 2.1 Searching analogous problems

These activities will now be briefly introduced in the context of our application.

In the early training steps the main concepts and methods relevant to the domain are acquired. At this stage, training exercises are used to build the initial know-how through direct the application of definitions and theorems. Then both exercises and the knowledge of the domain evolve gradually to more complexity as more deductive steps are used in resolutions. In the same time the human learner assimilates this knowledge and his skill is improved on the basis of cumulated experience. His ability to solve problems depends on his capacity to structure past experience, store it at long term and eventually utilize it properly. Whenever a new problem is submitted to the Work Memory (WM) (see Fig. 2.1), analyzing it and detecting some relevant evidence may evoke a similar situation already "assimilated" in the Long Term Memory (LTM). If an analogical correspondence is *checked* then some previously stored resolution plans will be used to solve the new problem [CAU.85].

We now introduce a model of case-based reasoning – a paradigm of particular reasoning by analogy – that makes it possible to solve problems in geometry on the basis of past experience assimilated on the basis of knowledge learned symbolically. Further we indicate how the LTM has been designed and how it evolves along time.

3. A Model of Case-Based Reasoning

3.1 Definitions

Let A and B be two sets of propositions on the same knowledge domain. These propositions represent descriptive elements (names, attributes) and relations (properties, functions, structures) of objects of the domain.

A (resp. B) contains two subsets of propositions h(A) and c(A) (resp. h(B) and c(B)). Assume that h(A) and h(B) are known (*hypotheses*) while c(A) and c(B) depend on h(A) and h(B) (resp.) through some dependency relation (formally introduced in Sect. 3.a.1°). Further assume that c(A) is already known or proved while c(B) needs to be proved (these are *conclusions*). We may say that h(A) and c(A) belong to a *source situation* while h(B) and c(B) belong to a *target situation*.

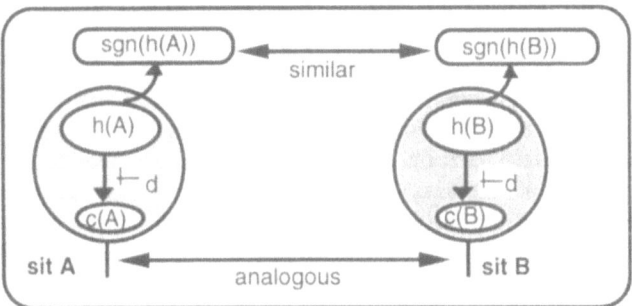

Fig. 3.1 The analogic model

The **analogic model** that we use is based on the classical paradigm [CHO.82, 86]:
   "c(A) *relates to* h(A)" **in the same way** "c(B) *relates to* h(B)".
   It comprises:

4   E. Chouraqui and C. Inghilterra

3.1.1 The Dependence Relation

The link between subsets h(A) and c(A) is a *dependence relation* that we denote:
⊢—d .
We write: h(A) ⊢—d c(A) and   h(B) ⊢—d c(B) .

This dependence relation ⊢—d , belongs to the common domain of situations
A and B. It is function of some relevant characteristics of subset h(A) or h(B).

Now we may denote situations A and B as follows:

  sit(A) = {h(A), ⊢—d , c(A)} and   sit(B) = {h(B), ⊢—d ,c(B)}

Let us assume that sit(A) and sit(B) are *similar*. "Similarity", here, is a matter
of resemblance between several characteristic features or relations.

3.1.2 Signatures

In a given knowledge domain we may select a list of propositions $P = (p_1, p_2,$
$..., p_n)$ relevant to a particular *point of view*. Here, the description point of view
is made of surface (descriptive) features and structural (relational) features
pointing at the methods for solving a goal [NGU.90].

Using these properties collected in vector P, each h(A) (resp. h(B)) may be
mapped to a vector

  $(a_1, a_2, ... , a_n)$ (resp. $(b_1, b_2, ... , b_n)$) belonging to $\mathbb{N}^n$.

This mapping is called a *signature* of sit(A) (resp. sit(B)):

  $sgn(h(A)) = (a_1, a_2, ... , a_n)$ and $sgn(h(B)) = (b_1, b_2, ... , b_n)$

Intuitively, sgn(h(A)) (resp. sgn(h(B)) are characteristic abstracts that will
serve as a basis for assessing similarity as unambiguously as possible.

Each coordinate a_i (resp. b_i, with i = 1 to n) is computed on the basis of
property p_i (belonging to P) relevant to the domain (see Fig. 3.2).

To compute signature coordinates we proceed as follows:

  • Let $P = (p_1, p_2, ... , p_n)$ be the set of relevant properties of the domain.
    Each property p_i is assigned a weight n_i measuring its importance.
  • **For each** i = 1 to n:
      if   at least one occurrence of each property p_i is found in h(A)
          (resp. h(B)),
      then assign coordinate a_i (resp. b_i) of signature $(a_1, a_2, ..., a_n)$
          (resp. $(b_1, b_2, ... , b_n)$), weight n_i associated to property p_i,
      else coordinate a_i (resp. b_i) is null.

The weighting of coefficients is determined according to the chosen 'sgn'
point of view. The aim is to highlight the salient and comparable features of the
similarity in order to "measure" it.

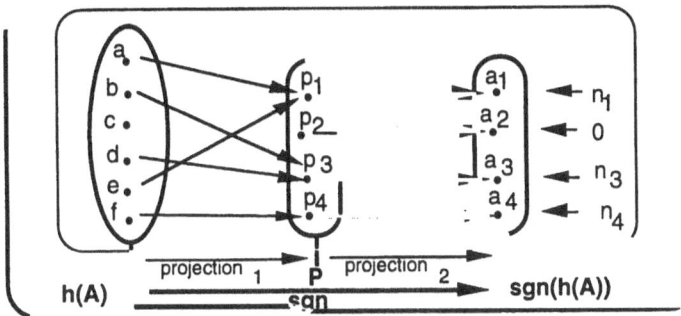

Fig. 3.2 The signature vector

In terms of symbolic machine-learning [MIC.83], list **P** and signatures (see Fig. 3.2) may be formally interpreted as follows:
   • *List P* is obtained through an inductive generalization of properties contained in subsets $h(S_k)$ of situations, i.e. by suppressing some conjunctive terms and transforming constants to variables in those properties. This is illustrated by projection1 (see Fig.3.2): For every $X_1, X_2, ..., X_n$, $P = (p_1(X_1), p_2(X_2), ... , p_n(X_n))$,
   • *Signature* is a *specialization* of list **P** in given situation S_k. Quantifying coordinates p_i in **P** yields vector $(k_1, k_2, ... , k_n)$ of $\mathbb{N}^n$. This is illustrated by projection2.

3.1.3 The Similarity Relation

Signatures make it possible to define similarity. A *similarity graph* representing the similarity between $sgn(h(A))$ and $sgn(h(B))$ can be constructed as follows:
   For every i from 1 to n: (a_i, b_i) is an edge of the similarity graph
   iff $a_i \neq 0$ and $b_i \neq 0$
The set of edges (a_i, b_i) denotes properties p_i common to h(A) and h(B), i.e. prolegomenae of the similarity evaluation.

3.1.4 The Similarity Ratio

The *similarity ratio* sim(A, B) between A and B is obtained as follows:

$$sim(A, B) = \frac{\sum_i min(a_i, b_i)}{max(\sum_i a_i, \sum_i b_i)}$$

where: (a_i, b_i) is an edge of the similarity graph, and $i = 1...n$

This ratio indicates the proportion of properties common to the two signature vectors. Its value belongs to interval [0,1] (see example in Sect. 4.2.4).

3.1.5 The Analogic Relation

If ratio sim(A, B) is greater or equal to some *threshold value* a, situations A and B are assessed as *analogous*. We denote: **analogous (A, B)** ⇔ **sim(A, B)** ≥ **a**

The threshold value is an estimate of the minimum proportion of common relevant properties ensuring that the analogical process will operate in a plausible way.

3.2 Reasoning by Analogy

Reasoning by analogy here amounts to asserting conclusion c(B)

given sit(A) = { h(A), ⊢—d, c(A) } and h(B).

Since sit(A) is in the LTM and sit(B) in the WM, the necessary steps of this reasoning are:

1°/ Search LTM for a source situation

sit(A) = { h(A), ⊢—d, c(A) } such that: ⊢— analogous(A,B) ,

2°/ Transfer dependence relation ⊢—d into target sit(B),

3°/ Use relation ⊢—d to try asserting c(B) from h(B).

In terms of symbolic machine-learning, we may use *generalization rules* [MIC.83], to formalize this process as follows:

• *Given:* h(A) and c(A), *we generalize* h(A) and c(A) *as* For every X h(X) and c(X) ,

i.e. replacing constant A with variable X,

and For every X h(X) and c(X) *as* For every X c(X) ,

i.e. suppressing a conjunctive term,

and For every X c(X) *as* For every X not h(X) or c(X),

i.e. appending a disjunctive term, which may be written equivalently:

For every X: h(X) => c(X) ,

• *Since :* ⊢— analogous (A,B) , X *is instantiated as* B ,

so that the preceding relation may be written: h(B) => c(B) ,

• *Besides,* ⊢— h(B) ,

• *Therefore we obtain*: ⊢— c(B) *by modus ponens.*

This reasoning is not entirely deductive since the generalization step is inductive and requires formal precautions [DAV 87]. Therefore, the issue of justifying analogy is partly linked with that of validating generalization from a single instance (here, A) [RUS.87]. It is known [MIC.83] that the likelihood of inductive generalization is a function of the number of instances verifying it and, above all, of initial constraints conditioning it and restricting the production of inductive hypotheses [CHO.82,86]. Because of this we supposed from the start that situations were similar, a hypothesis that is reinforced by the strong analogic condition (see the *threshold value*).

Nevertheless, inferences obtained in this way are not always valid, so that analogical reasoning may not be seen as a demonstrative reasoning. As rightly put by POLYA, reasoning by analogy is a heuristic type of reasoning *"... that one should not view as final and rigourous, but simply as temporary and plausible, aimed at finding the solution of the problem under investigation"* [POL.62]. It is equally absurd to reject reasoning by analogy and to endow it with unverifiable claims.

3.3 Analogic Inference

Keeping in mind the preceding remarks we may formalize *analogic inference* as follows:

- *Given* sit(A) = { h(A), ⊢d, c(A) } such that: h(A) ⊢d c(A) ,
- *It may be asserted*: For every sit(X): analogous(A, X) =>
    (h(X) ⊢d c(X) *is plausible),*
- *And, since:* sit(B) is such that ⊢ analogous(A, B)
    X *may be   instantiated as* B ,
- *Therefore:* h(B) ⊢d c(B) *is plausible* by *modus ponens.*

We view the following propositions as the Fundamental Hypothesis of the Analogic Model:

> - Given sit(A) = {h(A), ⊢d, c(A)} such that: h(A) ⊢d c(A),
>   - For every sit(X) = {h(X), c(X)}:
> ─────────────────────────────────────
> - analogous(A, X) ⇒ (h(X) ⊢d c(X) *is plausible*)

To render this model operational we need to sketch out the Analogous Syllogism:

For every sit(X) : analogous(A, X) ⇒ (h(X) ⊢d c(X) *is plausible*)
Since: ⊢ analogous(A, B)
───────────────────────────────────────
Then :   h(B) ⊢d c(B) *is plausible .*

4. An Application to the Resolution of Problems in Geometry

In this section, a method for solving problems of elementary geometry using analogic heuristics is introduced. Examples are borrowed from [BAR.83].

4.1 Problem Solving

4.1.1 Definitions

Let us first define how we view problem solving in geometry or mathematics [CUP.88]. This activity is normally based on two phases – reasoning and proof – that may be consecutive or interwoven (serial/parallel) [GUI.90]:

– Reasoning manipulates known facts (objects and relations) to generate new facts. To this purpose it makes use of theorems that are selected on the basis of heuristic strategies (for instance, analogy). Using inference rules it applies these theorems on facts to produce new facts. In other words, reasoning amounts to constructing a *planning network* determined by heuristic choices.

– Proof makes use of this planning network to construct a *deductive network*. This network is a sequence of formal hypothetico-deductive steps proving the goal under investigation. This sequence will be validated through a pedagogical control.

4.1.2 Formalization

Let B be a problem that needs to be solved using hypotheses $h(B)$, in which conclusion $c(B)$ should be proved on the basis of dependence relation $\vdash_d$.

Solving B amounts to reasoning (**Reas(B)**) and proving (**Proof(B)**) on B.

— As to **reasoning** *(i.e .planning network)* we may write:

$$\textbf{Reas(B)} = (< h_1(B), s_1, t_1, r_1, c_1(B) >, \; ..., \; < h_i(B), s_i, t_i, r_i, c_i(B) >,$$
$$...$$
$$..., \; < h_n(B), s_n, t_n, r_n, c_n(B) >)$$

with:

- $h_1(B) = h(B)$; $h_i(B) = h_{i-1}(B) \cup \{ c_{i-1}(B) \}$ and $i = 2$ to n,
- $c_n(B) = c(B)$: this is the stop condition of the reasoning: the goal is proved;
- s_i: is a heuristic strategy (e.g. analogy) that will be used to "explain" why theorem t_i was used at this step;
- t_i is a theorem whose premises are in $h_i(B)$ and conclusion is $c_i(B)$;
- r_i is an inference rule.

— As to the **proof** *(i.e. deductive network)*:

$$\textbf{Proof(B)} = (< h'_1(B), t_1, r_1, c_1(B) , p_1 >, \; ...,$$
$$< h'_i(B), t_i, r_i, c_i(B), p_i >, \qquad ...,$$
$$< h'_n(B), t_n, r_n, c_n(B), p_n >)$$

with:

- $h'_1(B) \subset h(B)$, $h'_i(B) \subset h(B) \cup \{ c_1(B), ..., c_{i-1}(B) \}$ and $i = 2$ to n ,
- $c_n(B) = c(B)$: the end of the proof – the goal is proved;
- t_i is a theorem whose premises are $h'_i(B)$ and conclusion is $c_i(B)$;
- r_i is an inference rule ;
- p_i is the formal proof of the i-th step of the demonstration.

The *heuristic aspect* takes care of the reasoning phase while *formal procedures* take case of the proof phase.

The advantage of dissociating the two processes is to bring into evidence their main components and to indicate precisely where they become operative: strategies (such as analogy) and explanations are used in **Reas(B)**, while deductions and proofs (justifications) are used in **Proof(B)**.

In this way it is easy to use them properly in the tutoring phase – help or control.

4.2 Defining Analogous Problems

Situations A and B of the model are problems of geometry in which h(A) and h(B) are hypotheses, i.e. a conjunction of properties of geometric elements (points, straight lines, segments,...) and c(A), c(B) are questions, i.e. goals or conclusions that need to be proved. These questions are also properties of the same kind , e.g. prove that "some point is the middle of some segment".

4.2.1 The Dependence Relation

If   ⊢— denotes the *deduction relation* of classical logic,
then $h(A) ⊢— c(A)$ is true,
if there exists a sequence of properties:
   $< h(A), c_1(A), \dots , c_i(A), \dots, c_h(A), c(A) >$
   produced by applying a set $d = \{ t_1, t_2, \dots , t_k \}$ of theorems of
   geometry [THA.88].

We define the dependence relation, denoted ⊢—$_d$, between $h(A)$ and $c(A)$ as follows:

• $h(A)$ ⊢—$_d$ $c(A)$ is true *iff* there exists a set of theorems
$d = \{ t_1, t_2, \dots, t_k \}$, produced from $h(A)$ by deduction $c(A)$,
thereby meaning that: $h(A)$ ⊢— $c(A)$ is true.

4.2.2 Constructing the List of Properties

The *description point of view* should at first highlight the characteristic features attached to the *resolution* of the goal under investigation. First, we build **P** = $(p_1, p_2, \dots , p_n)$, the list of characteristic properties estimated necessary for a proper assessment and evaluation of similarity. This list contains "basic" figures with their attached properties, e.g:

P = (triangle, right-angled-triangle, inscribed-right-angled-triangle, circle, parallelogram, diamond, *adjacent-segments, equal-length-segments, middle, perpendicular, parallel, cord, diameter*).

This list is bound to the general structure of the problem rather than to a particular goal. Actually, in the course of proving a goal different sub-goals may be defined that do not necessarily invoke the same theorems.

4.2.3 Calculating Signatures

The next step is to assign weights to signature coordinates. First it should be noticed that the facts needed to instantiate a theorem premises may be found (1) explicitly in hypotheses ; (2) in new facts extracted (derived) from hypotheses.
   To assign weights to signature coordinates a_i (b_i), indexes are matched with elements p_i of list **P** as follows:

> 1°/. **Extracting facts**:
> In all cases signature computation starts with a one-step deduction extracting the "almost evident" facts from all hypotheses.
>
> 2°/. **Assigning weights**: For every i = 1 to n:
> *If*   property p$_i$ of **P** is found in all hypotheses,
> *then*   coordinate a$_i$ is assigned coefficient 2,
> *else*   *if* it is *extracted*, *then* assign 1,
>   *else* if is not found, therefore assign 0.

A single occurrence of each property is taken into consideration. Repetitions would not increase the applicability of theorems under consideration ; beside, they would introduce errors in similarity ratios.

List **P** could be adapted to a particular goal if coefficients (properties) relative to this goal were assigned larger weights. However, since each goal contains sub-goals, in the end the whole list may become saturated.

4.2.4 The Threshold Value

The threshold value has been fixed to 0.7 on the basis of observations pertaining to several cases of analogy.

4.3 Commentary

The method outlined above has been obtained by processing a number of school problems [BAR.83] categorized into classes of analogous problems using the criteria defined above.

– Classes obtained in this way are "consistent" with respect to the sets of applicable theorems. Compared with classical sequential search methods, our method permits a better focussing on those theorems that are relevant to a given goal. To achieve this, the very first elements of list **P** (see Sect. 4.2.2) denote basic figures while the following ones denote properties common to all figures. In this way, signatures take in "figure prototypes" [GUI.90], a concept that has proved useful for finding goal resolution procedures.

– As far as weights are concerned (why take 2 or 1?), properties that are assigned weight 2 are the ones estimated most useful for proving the goal (and/or its sub-goals) ; properties with weight 1 do not necessarily come into consideration in the deductive network as they have been *systematically* extracted from hypotheses in a deduction step. Indeed, properties with weight 2 occur more often in deductive sequences.

– With threshold under 0.7 (i.e., sim(A, B) $\geq$ 0.7) the analogic heuristic becomes too unlikely. This value has been fixed on the basis of the sample set.

4.4 Examples

Let us illustrate the method with three problems [BAR.83]:

• **Prob. A.** Consider triangle ABC, perpendicular [AH], middle I of [AB], middle J of [AC]; (IJ) crosses (AH) in M. Prove that straight line (IJ) is the median of [AH] .

• **Prob. B.** Consider parallelogram PQRS with (PR) \ (QR), K is the middle of [PQ], F is the middle of [SR] and (KF) crosses (PR) in T. Prove that (KF) is the median of [PR] .

• **Prob. C.** Consider circle (C) with centre O, diameter [AB], M belonging to (C), (Bx) // (OM) and (Bx) crosses (AM) in P. Prove that (BM) is the median of [AP].

Using list **P** defined above, signatures of these problems are:

sgn(**A**) = (2, 1, 0, 0, 0, 0, 2, 1, 2, 2, 1, 0, 0),
sgn(**B**) = (1, 1, 0, 0, 2, 0, 1, 1, 2, 2, 1, 0, 0),
sgn(**C**) = (1, 1, 1, 2, 0, 0, 1, 1, 1, 1, 2, 0, 2).

Similarity ratios are: sim(**A, B**) = 0.82 and sim(**A, C**) = 0.54.

Therefore: ⊢— analogous (**A, B**) and ⊢— non(analogous (**A, C**)).

4.5 Analogic Inference

Suppose that problem prob(A) = { h(A), ⊢—d, c(A) } has already been solved and is stored in the LTM, with h(A) ⊢—d c(A) and d = { t1, t2, ... , tk }.

Solving problem B (for which only hypotheses h(B) are known) by analogy amounts to operate the following analogic inference:

> *For every* problem prob(X):
> analogous(A, X) ⇒ (h(X) ⊢—d c(X) *is plausible*)
> given d = { t1,t2, ... , tk },
> *Since :* ⊢— analogous(A, B) ,
> _____
> *Then*:   h(B) ⊢—d c(B) *is plausible* .

5. Memorization and Machine-Learning

Case-based reasoning by analogy makes use of information stored in the LTM; as a result, this information is enhanced so that the efficiency of future resolutions is increased.

5.1 Long Term Memory Architecture

The LTM contains the set of problems previously solved prob(A_i) = { h(A_i),⊢— d_i, c(A_i) }. These are represented as *frames* generating *instances* (Fig. 5.1). These frames are structured as a hierarchic network. The structural links are: *kind_of* (denoting a specialization to sub-frames) and *is_a* (denoting the actual realization of a frame in an instance). The root-frame "Solved_Problem" contains all attributes and procedures common to the frames it engenders.

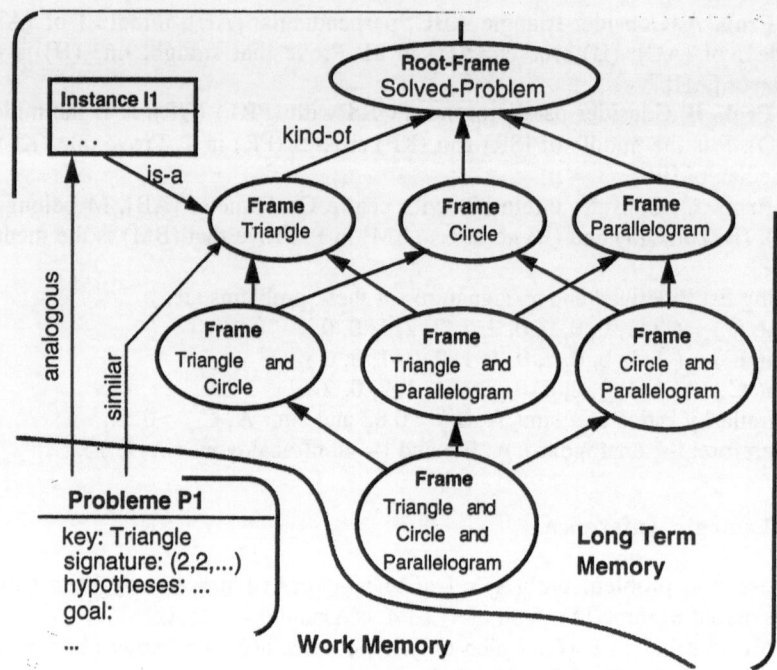

Fig. 5.1 Memory Organization

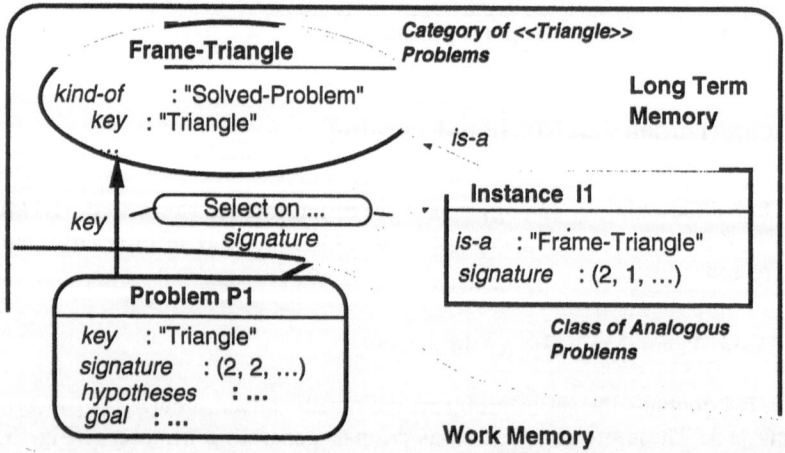

Fig. 5.2 Frame - Instance - Problem

• **How can one find the instance of an analogous problem already solved?**

Let P1 be a problem under consideration stored in the work memory. We use two search criteria, *key* and *signature*, to find another problem analogous to P1 in the LTM (see Fig. 5.2).

– *Key*: is an attribute intrinsic to the problem. It indicates what the basic figures (triangle, circle, ...) are found in its statement. Each frame represents a category of problems characterized by the basic figures used in each problem. This typology results from the information stored in the first elements of the list of properties **P**.

– *Signature*: each frame generates instances representing *classes of analogous problems*. Suppose for instance that a first problem P_1, solved in a classical way is stored into the LTM as instance I_1 of a frame S_C which it was pointing at (having the same key). For every new problem P_2 pointing at S_C we compute the similarity ratio sim(P_1, P_2). *If* sim$(P_1, P_2) \geq 0.7$, *then* instance I_1 "assimilates" P_2, *otherwise* P_2 is integrated as a new instance I_2 of frame Sc.

As a result, given two arbitrary instances I_1 and I_2 of the same frame S_C produced by problems P_1 and P_2, it can be asserted that

$$I_1 \neq I_2 \iff \text{sim}(P_1, P_2) < 0.7.$$

Therefore, non-analogous problems yield distinct instances.

• **Abstract:**

To select an analogous instance, two messages are send: the first message will use the key to search the category of similar problems (i.e., a frame); the second message uses the signature to evaluate the similarity ratio yielding the class of analogous problems (i.e., an instance of this frame).

Problems previously solved prob(A_i)= { $h(A_i)$, ⊢d_i, $c(A_i)$ } are therefore stored in instances. We will now examine how the relevant information is stored.

5.2 The "Solved_Problem" Frame:

Each frame contains all the knowledge needed for future resolutions. Fields (or attributes) have been defined so that they may enable Structuration, Selection, Storage and Evolution (see Fig. 5.3).

5.3 Storage and Evolution of the Memory Content

• **Memory evolution**

Memory evolves according to the following process:
  • Let B be a new problem in the work memory.

14   E. Chouraqui and C. Inghilterra

FRAME: Solved_Problem

% .... **Structuration attributes** ....
kind_of : list of FRAMES
% These are mother frames supplying heritages %
specialization : reference_value: list of FRAMES
% These are daughter frames specializing mother frames %
% .... **Selection attributes** ....
external_name : value: String
% This is the index or name of the first problem integrated %
key : reference_value: a FRAME
% These are basic figures, e.g. "Triangle_and_Circle" %
signature : value: (a_1, a_2, ... , a_n)
% This is a characteristic vector of the problem %
% .... **Storage and Evolution attributes** ....
   did_assimilate : value: list. < problem_#k, similarity_ratio >
 cardinality: < 0, ∞ >
 domain: similarity_ratio ≥ 0.7
 if_created: Assign1_Procedure
% Keeps the trace of extracted expertise. This procedure modifies the next field %
goal_sequences : value: list. < $goal_i$, $sequence\_list_i$ >
 cardinality: < 0, ∞ >
 if_created: Assign2_Procedure
% Contains planning and deductive networks (traces of resolutions).
This procedure modifies the next field once it has transformed sequence_list %
resolution_plans : value: list. < $goal_i$, $resolution\_pile_i$ >
 cardinality: < 0, ∞ >
% Below are theorems t_i belonging to set "d" of relation $\vdash_d$.
These are compiled as a list of resolution procedure piles used as a
"resolution toolbox"%
... etc.

Fig. 5.3 The Solved_Problem Frame

If every instance sit A_i of the LTM is such that: $\vdash$ non(analogous(A_i, B)),
then problem B is solved in a classical way
      and prob(B) = { h(B), $\vdash_d$, c(B) } is *integrated* into the LTM;
else there exists an instance prob(A_i) of the LTM such that:
      $\vdash$ analogous(A_i, B)
and *If*   h(B) $\vdash_d$ c(B) is true, *then*   prob(A_i) *assimilates* prob(B).
      else prob(B) is solved in a classical way, yielding a new relation
      $\vdash_{d'}$, so that
      instance prob(A_i) may be updated replacing dependence relation $\vdash_d$
      with $\vdash_{(d \cup d')}$.

• **Initialization – Integration**

Solving problems in geometry by analogy is only possible if the long term memory already contains instances $prob(A_i) = \{h(A_i), \vdash\!\!-d_i, c(A_i)\}$, each of which represents a class of analogous problems. This initialization is performed by a classical demonstration of the first problems submitted to the system. These problems are therefore integrated as instances of frames sharing the same key. (Remember that non analogous problems yield distinct instances, cf. Sect. 5.a).

During this integration instantiation, an attribute *goal_sequences* summarizes plan and deductive networks, and another attribute *resolution_plans* builds tables of frequencies of occurrence of theorems in the steps of the proof.

• **Updating – Assimilating**

Attributes *goal_sequences* and *resolution_plans* "assimilate" new knowledge, storing it into new resolution networks (see Sect. 4). Frequency tables cumulate new occurrences of theorems used in proofs.

6. Conclusion

Case-based reasoning enables the tutoring system to solve new problems on the basis of assimilated experience. Further it is able to remember the expert knowledge acquired from solving a problem. The way this tutoring system acquires expert knowledge discloses its capability of machine-learning at two levels: (1) refining the dependence relation on the basis of acquired expert knowledge, and (2) integrating new situations that may eventually become source situations themselves. Reasoning and machine-learning in this system will make it easier to construct the Student Model and the Tutoring Module. Knowledge imbedded in the tutoring system increases in the same way as the learner's knowledge. This synchronic resemblance should improve their mutual adaptability, hence their efficiency.

References

Bareil H., Zehren C. (1983) Mathématique 4ème. Hachette, Paris.

Bledsoe W.W. (1986) Some thoughts on proof discovery. in Proc. of I.E.E. symposium on Logic Programming. Salt Lake City, pp. 2-10.

Cauzinille-Marmeche E., Mathieu J., Weil-Barais A. (1985) Raisonnement Analogique et Résolution de Problèmes. L'Année Psychologique 1985, pp. 49-72.

Chouraqui E. (1982) Construction of a model for reasonning by analogy. Proceedings of E.C.A.I., Orsay, pp. 48-53.

Chouraqui E. (1986) Le raisonnement Analogique: Sa problématique, ses Applications, Journées Nationales sur l'I.A., Aix-les-Bains, pp. 107-117.

Chouraqui E., Inghilterra C. (1987) Modélisation des connaissances d'un système expert d'E.A.O. de la géométrie, ECOOP 87.Paris, pp. 175-180.

Cuppens R. (1988) La Résolution de Problèmes Mathématiques par un Homme ou une Machine. Actes de la 2° Université d'Eté: Intelligence Artificielle et Enseignement des Mathématiques. 4-8/7/1988. Publications IREM de Toulouse. pp. 85-110.

Davies TR., Russel SJ. (1987) A logical approach to reasoning by analogy. Proc. of the 9th Inter. Joint conf. on Artificial Intelligence, Milano, pp. 264-270.

Guin D. (1990) Modélisation des Connaissances pour un Système d'Aide à la Démonstration Géométrique. Deuxième Congrés Européen: "Intelligence Artificielle et Formation." APPLICA'.90 Lille 24-25 Sept.1990.

Inghilterra C. (1992) Apport de la représentation orientée objet et du raisonnement analogique dans la conception d'un tutoriel de géométrie. Thèse de Doctorat de l'Université d'Aix-Marseille III - spécialité Informatique, 7 juillet 1992.

Michalski R.S. (1983) A Theory and Methodology of Inductive Learning. In Michalski R.S.& al. (Eds). (1983) Machine Learning, an Artificial Intelligence Approach. Volume 1. Tioga Publishing Company. Palo Alto. pp. 83-134.

Nguyen-Xuan A. (1990) Le Raisonnement par Analogie. In Richard J-F., Bonnet C., Ghiglione R. (Eds). Traité de Psychologie Cognitive. Tome 2: Le Traitement de l'Information Symbolique. Dunod. Paris. pp. 145-157.

Polya G. (1962) Comment Poser et Résoudre un Problème. Ed. Dunod. Paris.

Pintado M. (1991) Apprentissage par Analyse des Solutions. Application à la géométrie élémentaire. In Proceedings of K.M.E.T. Sophia-Antipolis. April 22-24. 1991. pp. 149-162. Herin-Aime D. (Ed.). I.O.S. Press. Amsterdam.

Russel S.J. (1987) Analogy and Single-Instance Generalisation. In Proceedings of the Fourth International Workshop on Machine Learning. June 22-25, 1987. University of California. Irvine.

Thayse A. & al. (1988) Approche Logique de l'Intelligence Artificielle. Tome 1: De la logique classique à la programmation logique. Dunod. Paris.

Modelling Children's Informal Arithmetic Strategies

Roshni Devi, Tim O'Shea, Sara Hennessy, Ronnie Singer

Centre for Information Technology in Education
The Open University, Walton Hall, Milton Keynes MK7 6AA, UK

Abstract. This chapter discusses our approach to modelling, supporting and developing children's informal arithmetic strategies. The first section of the chapter is an account of a study that was carried out to investigate the development of the arithmetic concepts commutativity and associativity. The implementation of production-rule models of children's problem-solving strategies is discussed. The last section describes our computer-based package called Shopping on Mars, and reports on some field trials of the package.

1. Introduction

This chapter discusses our approach to modelling, supporting and developing children's informal arithmetic strategies. There are four main influences on our work. Hennessy (1986) has shown the value of children's informal arithmetic strategies and that it is possible to identify their use by individual children and track their development. Young and O'Shea (1982) have demonstrated that some aspects of formal arithmetical skills can be usefully modelled using production rule systems which can in turn be used as components of intelligent tutoring systems like that developed by O'Shea (1979) for quadratic equations. Smith (1986) in his innovative work with ARK (The Alternate Reality Kit) has developed an approach to interface design based on the uniform physical metaphor and direct manipulation which makes it possible for pupils to manipulate abstract concepts in science (e.g., O'Shea and Smith, 1987) and mathematics. Finally the work of the Open University in developing computer-assisted learning packages (reviewed in Jones, Scanlon & O'Shea, 1987) has shown the value and necessity of testing and refining packages with real users. Our approach to informal arithmetic involves longitudinal studies of children learning in classroom settings so as to identify both inter-pupil and intra-pupil learning effects.

Then we build computational models of pupil competence at different stages of development in order to clarify what exactly is being learnt and why pupil behaviour is changing. We subsequently construct and test computer packages which draw on these models for their instructional interventions and which incorporate direct manipulation interfaces for ease of pupil use. Finally we test our computer packages (and thus our prior analysis of the learning process) in a variety of school settings.

In the next section of this chapter we describe a study of the acquisition of commutativity and associativity. This study is representative of our empirical approach to understanding children's use of mathematical concepts and strategies. Section 3 is a brief discussion on the production-rule formalism for constructing models to account for pupils' observed performances. In section 4 we describe aspects of the evaluation of a package called Shopping on Mars. Shopping on Mars is representative in terms of instructional strategy, interface and mode of refinement of the types of computer-assisted learning packages we are developing.

2. A Study of Acquisition of Commutativity and Associativity

A study was carried out to investigate the development of the concepts of commutativity and associativity. The use of the concepts by children between the ages of 5 and 12 years was examined. The aims of the study were:

i) to identify the stages that children go through in acquiring the concept of commutativity for addition of integers.

ii) to achieve some understanding of why children generalize commutativity to subtraction.

iii) to gain insight into children's progression from commutativity to associativity.

2.1 Method

2.1.1 Subjects

The subjects comprised 105 children (49 boys and 56 girls) between the ages of 5 and 12. The sample consisted of four 5-to-6-year-olds, thirty-one 6-to-7-year-olds, thirty-three 7-to-8-year-olds, twenty-two 8-to-9-year-olds, eight 9-to-10-year-olds, five 10-to-11-year-olds and two 11-to-12-year-olds. Details of each subject (age, ability) are listed in Appendix 1.

2.1.2 Materials

The children were given a set of commuted pairs of addition problems like $4 + 5$ and $5 + 4$. The numbers in the problems were randomly selected. The problems were written on paper, one at a time, and read out. After doing the set of addition problems, a pair of commuted subtraction problems (e.g., $7 - 4$ and $4 - 7$) was also given. The three-numbered problems were of the type $5 + 8 + 5$. The children used fingers for counting. Some of them used rulers. Some ice-lolly sticks were also provided. A tape recorder was used to record the interviews.

2.1.3 Procedure

Children were given arithmetic problems (most of them being commuted pairs, i.e., x + y followed by y + x) to solve. They were interviewed to get details of their skills and strategies and to find out their levels of understanding of the concepts. The work was carried out in two stages:

    i) study of commutativity

    ii) study of transfer of knowledge of commutativity to solving 3-numbered problems.

To identify the different levels of understanding of the concept of commutativity, children at different stages/levels were studied. After they had written down the answer for the first of a commuted pair of problems, they were asked "Now, can you tell me if this will add up to x (where x is the child's answer) – the same as or different ...", "Why do you think it will be the same?" (or different, depending on their response). In addition, a pair of wrongly-answered, large-numbered problems (like 1023 + 4970 = 5985 for elder children, and 130 + 485 = 550 for the younger ones) were written down, read out loud by the experimenter, and then the children were asked "what will 4970 + 1023 (with the appropriate numbers) be?" The children who showed knowledge of the concept were tested to see if they generalized it to all numbers: "If I swapped the two numbers around, will the answer always be the same?", "Even for very large numbers? (If their answer to the previous question was "yes")", "Do you know why that is?" Furthermore, it was noted whether the subjects computed the sums for the second of the commuted pairs of problems, or whether they copied the answers from the previous problems. Finally, they were given a pair of subtraction problems to find out if they generalized the concept to subtraction. Details of the observations and measures used for deciding what level a child was at are described in Section 2.2.

Of the students studied for the stages of development of commutativity, 77 were studied for their performance on three-term problems.

The sample size of 77 students was a result of ignoring those students (especially the younger ones) who were not competent enough for the study (for example, those who did not know what to do or how to proceed on the 3-numbered problems. Some of these students had explicitly stated that they did not know how to do these problems).

2.2 Results

2.2.1 The Performance Levels of Commutativity

The study revealed several levels of performance reflecting understanding of the concept of commutativity. There are the basic levels where children use the 'count on from the larger addend' (COL) strategy for addition, because it is a faster means of arriving at the answer, without showing any evidence of conceptual knowledge of commutativity. There are the fully-developed stages where knowledge of the concept is applied to more complex situations (e.g., application to 3-numbered problems and invention of informal procedures for solving problems). As a result of the the behaviour/performances of the subjects in the study, I propose the following levels in the development of commutativity (of addition of integers):

i) **Order-irrelevance principle** (Gelman, 1977) – while assigning tags to objects in a set, it does not matter which tag is assigned to which object.

ii) **Implicit knowledge** – children might possess knowledge of the concept, but it is implicit (they cannot articulate it – some might not have the language/vocabulary to describe it). In addition, their knowledge might be at a level which does not allow them to make use of it. Hence, children at this level might compute the answer to x + y after having done y + x, instead of copying the previous answer.

iii) **Commutativity** – the realization that a + b is equivalent to b + a (for all values of a and b), and the use of the concept. This may proceed in steps (not necessarily in this order):

    a) concrete examples only,
    b) small numbers only,
    c) abstraction,
    d) generalization to all numbers.

The progression here is quite complex. The four steps can all be interrelated, and hence, a child can be at one or more of them at any one time. There are several possibilities of progression:

    concrete small —> concrete large
    concrete small —> abstract small
    abstract small —> abstract large
    concrete large —> abstract large

iv) **The application of commutativity to problems more complex than 2-term addition**. There are several levels of application:

    a) to concrete examples only,
    b) to small numbers only,
    c) generalizing to large numbers and to abstract examples,
    d) inventing informal procedures based on the concept for solving
       more complex problems,
    e) 3-numbered problems.

At this level, like level iii, the different sub-levels of application may be interrelated.

Note that the above levels do not necessarily have a psychological or developmental status. They are just descriptions of performance. A child need not necessarily go from performance level i to iv in that order, nor does s/he necessarily go through each one of them. An attempt was made at categorizing the subjects in the study into one of the four performance levels. This was done using the tape-recorded protocols, the written problems and their answers and notes on the observations made during the interviews. The result for each subject is listed in Appendix 1. The details of the observations and measures used for deciding the levels are described below:

i) While counting out a set, children often recounted (for various reasons: e.g., to make sure, or because they made a mistake, or they lost track of their count or because the experimenter asked them to do so). When counting again, they did not necessarily assign the counting sticks the same tags (e.g., '1') as they did the previous time. One child, for example, picked up the sticks and put them in her hand as she counted them "1, 2, 3, 4, 5". When she recounted (she put them on the desk this time), she assigned the tag '1' to the stick to which she had assigned '5' previously. Although some children's behaviour revealed knowledge and use of this principle, this observation was not necessarily the criteria that was used in deciding a level i child. Instead, a subject who was not categorised into the other 3 levels, was automatically classed as having performance level i. (Note that this level was identified more as a logical prerequisite for commutativity than as a result of performance behaviour).

ii) This was the most difficult stage to identify. Performance on the commutativity problems only revealed that the subject had not reached stage iii (the difficulty was to identify whether s/he had knowledge of the concept or not).

Note that no distintinction has been made between commutativity and Baroody and Gannon's (1984) concept of protocommutativity (the order in which addends are dealt with does not make a difference in terms of the correctness of the sum). Hence, children who might have had knowledge of protocommutativity only, have been included in this category. Children who only have a concept of protocommutativity, do not realize that the result of $3 + 5$ is the same as that of $5 + 3$. They know that they get the answer in either case – whether they start from the first addend or from the second. This behaviour was noticed in children's performance on concrete tasks as well as on abstract examples. On concrete tasks, children at this stage, were observed not paying particular attention to the order of the addends, hence counting out either of the addends first. On abstract problems, there were several children whose level of understanding of commutativity was at this stage and who were using the COL strategy while adding. When asked why they had started from the larger addend, and not from the first addend, the replies were those that did not reveal knowledge of commutativity: for example, "because it is easier", "because it is faster", "because x is larger than y". When subjects at level ii were asked if the sums/results of a pair of commuted problems will be the same or not, they did not immediately say "same" (some started computing the answers). Even if they said "same", their explanations referred to the similarity of the two addends (which does not necessarily extend to their sums).

iii) It was common to find children at stage a) and at stage b) of this level. Due to the interrelationships between a, b, c and d, it is difficult (if not impossible) to be precise about which one of these stages a child is at. Hence, a child was classified as being at level iii if s/he was analysed as being at any one of these stages. A child was classified as being at this level if at least two of the following criteria were satisfied:

i) for the large-numbered, incorrectly-answered problem, s/he responded with the same answer.

ii) generalized to subtraction.

iii) used the answer from the previous problem for answering the second of a pair of commuted problems.

iv) if his/her explanations included statements which referred to the *sums* of two commuted problems being the same (including use of pronouns like *it*, *they* referring to sum(s)), e.g., "They are the same, but you have swapped them around and it equals the same number".

iv) A child was classed as being at level iv if s/he used grouping (did not carry out the sum from left to right) for solving the 3-numbered problems.

2.2.2 Generalization

A total of 63 students' performance on a pair of subtraction problems, like 4 − 2 and 2 − 4 (after being tested on commutativity of addition), was examined (The results of the individual children's answers are listed in Appendix 1). Their responses to the second of the pair of problems fell in one of the following categories:

i) "cannot be done" or "not possible",
ii) "zero",
iii) applied commutativity − copied the answer to the previous problem,
iv) subtracted the smaller number from the larger one.

The proportion of students' responses in each category is listed in Table 1.

Table 1. Numbers of students in each category

not possible	0	over-generalize	smaller from larger
24	20	15	4

Eight of the students in the categories '0', "over-generalization" and "smaller from the larger" originally thought that it "cannot be done", but thought that this was like giving up on a problem, and hence attempted to arrive at solutions.

2.2.3 Associativity

The subjects used one of the following strategies in solving 3-numbered problems (like 4 + 8 + 4):

i) Performed the operations in any order. They *grouped* any 2 of the 3 numbers, and carried out the appropriate operation on them first. Hence, in a problem like a + b + c, they did either a + b first or a + c first or b + c first (whichever was easier). For example, for 3 + 5 + 5, children using this strategy, did 5 + 5 first because most children know doubles, and/ or it is easier to add on a small number at the end and/ or 10 is easier to work with than 3 or 5.

ii) Performed the operations from left to right (linear strategy). That is, they always did the operation on the first 2 numbers first, and then performed the next operation on the result and the last term. Thus, $4 + 8 + 4$ would be solved as: $4 + 8 = 12$, $12 + 4 = 16$.

Table 2 below is a comparison of the above strategies used by children who used the concept of commutativity (stages iii and iv) and those who did not (stages i and ii). It shows that knowledge of commutativity and use of grouping are strongly related. Not knowing commutativity implies not using grouping. It also shows that knowing commutativity is a necessary but not sufficient condition for using grouping.

Table 2. Relationships between strategies for 3-numbered addition and knowledge of commutativity.

	strategy i (grouping)	strategy ii (linear)
Knew commutativity	50	11
Did not know commutativity	1	15

The results above show that of the 61 students who knew commutativity, 82% of them used grouping on the 3-numbered problems. These students also gave descriptions that showed explicit knowledge of associativity. For example, "You have just changed the order of the sum", "It doesn't matter what you do first, it's the same thing".

There were 3 children who did not apply associativity, but showed explicit knowledge of the concept. PKP and AN revealed this by not computing the answer to the second of these problems again:

$6 + 8 + 2 = 16$ $(8 + 6 = 14, + 2 = 16)$
$2 + 8 + 6 = 16$

When asked how they had done this, they replied "You repeated the question" and "it is the same as that one" (pointing to the first problem) respectively.

MSN: $3 + 2 + 3 = 8$ $(3 + 2 = 5, + 3 = 8)$. When asked for any other way of doing this problem, she replied "Can also do $3 + 3$ first and then add 2".

There were 3 children who knew the concept of commutativity, but did not extend it to 3-numbered problems. Their inability to transfer the knowledge of commutativity implies that there is a stage between knowledge of a concept and its application. If it was not for this stage (revealed by these 3 children), commutativity would be a necessary *and sufficient* condition for grouping.

An interview with the child (PD) who did not know commutativity, and yet seemed to apply associativity (see table 2), revealed that she did not have any knowledge of the latter concept either. She was using the labour saving strategy, COL, while doing the commutativity problems. For the problem, $7 + 13$, she started counting on from 13. Then, for $13 + 7$, she repeated the counting. When asked explicitly if she was aware that the order in which 2 numbers are added does not make a difference to their sum, she replied "No". On the 3-term problems, she began from the largest addend. For example, 3 + 3 + 12 - she started from 12 "because 12 is the largest number". This strategy seems to be an extension of the COL strategy that she used on the 2-term problems. This child's algorithms embody the concepts of commutativity and associativity, but she does not consciously/explicitly possess that knowledge. PD's behaviour highlights the need to distinguish between the knowledge of a concept and the use of algorithms that presuppose the concept. This type of distinction has also been discussed by other researchers (Baroody, 1984; Hennessy, 1986; Resnick, 1983).

2.3 Discussion

2.3.1 Levels of Performance of Commutativity

Different levels of performance of the concept of commutativity have been identified. However the complex interrelationships between the sub-levels need to be studied in more detail. It is hoped that a longitudinal study (which is being carried out at the moment) will help to refine and/or validate the performance levels.

In this study, commutativity of one operation only (addition) has been considered. One would expect variations in the results if the other operations are considered as well. Another dimension of variation would be the type of number system (integers, fractions, decimals, etc.).

2.3.2 Generalization

Table 1 shows that 70% of the 63 subjects who were tested on subtraction problems where the minuend was less than the subtrahend (e.g., 3 − 5), reasonably gave either "not possible" or '0' as their answers.

One of the reasons for children over-generalizing could be that they have been led to do so. They think that because the teacher has given the example with the commutativity examples (i.e., the teacher's aim is to teach commutativity), this must be an example of commutativity as well. In addition, generalization is a way of learning. With the positive examples of commutativity that they have seen so far, there is no reason for them not to believe that everything is commutative. From this, one can say that students' interpretation of the situation (how they think about it) and the type of examples are two possible variables responsible for their performance.

Another possible reason for generalization is that the children do not have a concept of negative numbers, which is the prerequisite knowledge for solving problems like 4 − 7 so they attempt a repair. One of these repairs happens to be applying commutativity.

The children who give '0' or "nothing" as answers do not all have the same reason for their answers. In reply to "Can you show me how you would subtract 4 from 2?", some of them revealed that there is nothing left to subtract from after subtracting 2 (of the 4) from 2; some explained that there is nothing left after subtracting 2 (of the 4) from 2 and hence the answer is 0. Most of these children reason like Su (6 years, 9 months):

Su: (*pause*) I took 4, I take away. No! I took 2 (showing 2 fingers), from that I (*pause*). I have to minus (*pause*). I took 2, from that I took 2 away; there's 0 left.

Another child who reasoned like Su exclaimed "None left" while trying to work out $2 - 4$ on her fingers. These children's answers are quite reasonable, since they realise that they cannot go beyond zero.

For some of the other children who say "nothing" or '0', it is another way of saying "I don't know" or "I don't know what's happening here". Yet other children mean to say "It cannot be done". This shows the importance of pinning down a child's meaning of his/her response, since it is not always clear what a response on a given task means.

2.3.3 Associativity

There were some children who had knowledge of the concept, but did not use it. Gelman and Gallistel (1978) provided empirical evidence of children who appeared to lack understanding of a concept on one task and showed performance consistent with the concept on another task. This implies that children do not always use their conceptual knowledge.

Some possible reasons for this are:

i) they do not think of the concept at the time. For example, in case of commutativity: "I knew it but I calculated it again because I didn't think of it".

ii) even if they are aware that a particular concept is applicable, they know that after all they will get the same answer whether they use it or not. For example, there were a number of children in the commutativity experiments who knew the concept and were computing the answers to the second of the commuted pairs of problems. They were doing this so that they could check their answers to the first problem. In general, children do not always use the most efficient strategy in their problem solving.

iii) they have not reached the stage in the development of the concept where they can apply the concept.

The studies suggest that the acquisition of associativity and that of commutativity are interrelated. There were instances of children who were at stage iii (for example using commutativity for small numbers only) and were using grouping on 3-term problems. There were also children who were noted for applying commutativity to 3-term problems (partially – only to the first 2 addends). There was not a single child who was noted for using grouping (commutativity or associativity) on 3-numbered problems and not using commutativity on 2-numbered problems. This shows that a complete understanding of associativity does follow from understanding of commutativity.

3. Production-rule Models

This section demonstrates the task of building computational models of pupil competence. The models are used in a tutoring system, like Shopping on Mars, for recognizing students' strategies.

Based on the results of the study reported above, production-rule models of children's strategies for solving arithmetic problems like $5 + 6$ and $8 + 5 + 9$ have been implemented. An individual's observed problem-solving strategies are represented by a set of fine-grained rules. A rule is of the form

>   if <problem state> then <perform action>

The results of the actions are stored in memory; the memory is updated after each action. The updated memory shows the stage in the problem-solving process that has been reached. The state of the memory is matched against the condition (problem state) sides of the production rules in order to select the next rule to 'fire'. At this stage, more than one rule could match the memory items. To select one of the possible rules, production systems have conflict resolution strategies.

The implementation uses the simple conflict resolution strategy, rule ordering, that is, it selects the first of the list of possible rules. In addition, a rule may match more than one memory item. To overcome this problem, the implementation chooses the item that was added last (most recent item) to the memory. (The last action triggers the next action). This cycle of 'firing' of rules to solve a problem using a particular strategy continues until an answer is reached.

For example, the following is a trace (actions) of a sequence of production rules that 'fire' for solving the problem $5 + 6 + 6$, using the grouping strategy:

>   Group-first 5 6 6
>   Lookup unsuccess 5 6
>   Group-second 5 6 6
>   Lookup success 12 5
>   Left-over 5
>   Part-answer 12
>   Count-on 12 5
>   Answer 17

Using the grouping strategy, the production system needs to decide which two of the three numbers to add first. To do this, it tries all the three combinations, looking up a database of known facts each time to find out if it knows the sum of the two numbers, until it succeeds. The second line of the trace, 'lookup unsuccess 5 6' means that the program does not have $5 + 6$ as a prestored known fact. The result of a successful 'lookup' is 'part-answer' of the two addends and the 'left-over' addend. These two numbers are then added using the 'count-on' strategy to get the final answer.

The next section is a description of a tutor that incorporates production rules for modelling children's informal arithmetic strategies. Strategies, which embody the concepts commutativity and associativity, identified in the study reported above, are typical of those modelled in Shopping on Mars.

4. Shopping on Mars

4.1 Design

The overall objective of the Intelligent Arithmetic Tutor project (described in more detail in Hennessy et al., 1989) is to help children learn to treat arithmetic as a means for manipulating representations of quantities and to help them learn to formulate shortcuts and efficient methods for problem solving.

The activity takes the form of an adventure game and is implemented on the Acorn Archimedes 310 microcomputer. It is aimed at children aged 8–12 and has minimal linguistic content. A range of skills and concepts in primary school arithmetic instruction are supported. The main purpose of the activity is to help children realise that different types of problems are sometimes best solved using different calculation methods and that the appropriateness of calculation techniques depends on problem structure as well as on the numbers involved.

The tasks presented in Shopping on Mars are as similar as possible to those encountered in everyday, 'real world' situations such as shopping. The problems involve addition, subtraction and multiplication of positive whole numbers. The children need to (a) decide which operation is to be performed in each problem situation, (b) choose a computation method and perform the calculation, and (c) report the result using a special language for representing mental algorithms on the computer.

4.1.1 The Game

The Shopping on Mars game proceeds as follows. Two players land on Mars in a rocket with no remaining fuel and proceed to cooperate in a series of purchasing tasks. The players' primary aim is to obtain 'fuel' with which to return to Earth. They are represented by astronaut figures who navigate around the Mars landscape in a 'Marsmobile' in search of the fuel shop. Their progress is thwarted by a series of obstacles which may be overcome by buying certain items at nearby shops. A chronological record is kept of how far individuals have progressed and of their difficulties. Student records are stored on personal disks for access by teachers and pupils. These records are used to construct the current model of the student whenever they use the activity.

4.1.2 Interface

The child has access to three basic views. The first is a global view of the Martian landscape showing the position of the Marsmobile relative to the rocket and Martian shops. External obstacles occur at this level. An example of an obstacle is a landslide which creates a rockfall, automatically cleared after a wheelbarrow and shovel have been purchased. Shops selling appropriate items are situated at various points along a road between the rocket landing site and the fuel shop. The order of the obstacles encountered and thus the route taken between shops in each game is altered for variety.

Where appropriate, screen images are zoomable so that close-up perspectives can be taken. Thus, the inside of a shop looms into view automatically upon reaching its door. There is a counter with a till and a sliding tray, a Martian shopkeeper behind the counter, a display of goods labelled with price tags, a shopper with money contained in a zoomable purse and a zoomable shopping bag (clicking on these items enlarges them). The bag is shared between the shoppers and acts as a container for their purchases. Items for sale may include a decoy item or a working calculator, which can be used twice before its batteries run out. The control panels show whose turn it is at any one time via colour-coded name labels; the active child's name is highlighted. Finally, when the shopper moves up to the counter, it zooms into close-up view so that maximum space is available for the purchasing transaction. The purse, calculator, money and objects to be bought can then be easily manipulated.

4.1.3 Playing the Game

Having encountered an obstacle, the shoppers drive to a shop in order to purchase an item with which to overcome the obstacle. Once inside the shop, the active shopper acts by clicking on objects it wishes to buy and taking them to the counter. Additional tasks entail finding the total cost of two items with different prices, subtraction involves calculating what a discounted item will cost, and multiplication involves calculating how much x items at y pence would cost. After calculating the total cost, the players have to hand over to the Martian shopkeeper the correct amount of money (from the zoomable purse), and check their change, if any.

Instructions and help are given to the children by the shopkeeper at various points during the game; they are spoken and simultaneously printed on the screen in a message box. In particular, children are encouraged to check their change after each transaction and to indicate that it is right or wrong. The shopkeeper sometimes shortchanges the child blatantly. Occasionally, players will be asked to demonstrate how they worked out the total cost; they then enter an account of their method using our description language (described in Sect. 4.1.5).

4.1.4 Instructional Architecture

The program's overall goal is to foster the development of a flexible repertoire of mental arithmetic algorithms, coupled with the ability to select the most appropriate one for a given problem. Appropriateness depends of course on the individual student; the tutor's control structure involves first ascertaining what calculation-specific concepts and methods the child already possesses, and then selecting a new method to teach the child. The method chosen is constrained by the child's understanding of arithmetic.

The program consists of a set of tutorial rules, which encode heuristic information about what action to take next: e.g., whether to teach, to model a child's problem-solving behaviour, or to alter the level of problem difficulty. The system's two other main components are a diagnostic module (for modelling calculation methods and inferring the conditions constraining the child's use of a rule), and a problem generator which can set up a problem for the child. It is not normally possible to ascertain what mental arithmetic algorithm the child has used on the basis of answer data alone; intermediate step information is indispensable. A special graphical description language was created to meet this need; it also serves to reduce considerably the huge space of rules which must be searched.

4.1.5 The Description Language

In developing a tutor for this domain, we were faced with the problem of designing a language for describing mental arithmetic which is rich enough to encompass the diverse processes used by children, sufficiently simple to be easily learned and used, and suitable for demonstrating new algorithms to the child (i.e., the language used for exemplar presentation should be the same as that used to obtain student protocols). To this end, we have developed the Graphical Arithmetic Description Language (GADL). It again embodies the principle of direct manipulation and is based on the simple metaphor that numbers and arithmetical expressions are objects which can be manipulated by arithmetical tools (Evertsz, Hennessy and Devi, 1988). The mouse operations and movement conventions are carried over from the 'shopping' interface.

The GADL interface consists of a number of objects, and six on-screen tools for manipulating them. Objects comprise encircled numerals and arithmetic operation symbols which can be hooked together (a 2 joined to a 6 forms 26). Tools include incrementing and decrementing tools (which take arguments), a deletion tool and a number fact one. There is also a 'partitioning tool' for splitting up a number or the terms of a problem. Both objects and tools are accessed via a mouse-driven grabbing hand. The screen is scrollable so that the child can access and return to previous steps. It is important to note that GADL is not a calculation device; the child has already solved the problem mentally and is now retrospectively describing what s/he did. Although this description often takes longer than the original calculation, the graphical representation reduces working memory load since all of the child's solution steps are explicitly available on the screen. This feature is just as important when the shopkeeper demonstrates solutions through the medium of GADL.

4.2 Field Trials

4.2.1 Interface

We have tested the Shopping on Mars program extensively in two local primary schools; so far, 120 children aged 8–12 have taken part in the field trials. The methods we have used comprised clinical interviews and protocol collection, coupled with videotaping and analysis of child-computer interactions. The collection of video data (in conjunction with an experimenter's note-taking) provided an effective and accurate record of the children's problem solving during testing and the children showed no adverse reaction whatsoever to being filmed. Most of the time, we recorded the screen image – in conjunction with the children's verbal interaction on the soundtrack – so that the detail of their work with the system could be analysed. This type of video was felt to be more valuable during development work than filming the children themselves.

In order to iron out at an early stage any obvious problems arising with the interface, a cardboard version of the game was used before programming on the Archimedes was completed. We constructed a simulated Martian shop, employing plastic money and cardboard materials. One experimenter acted as the shop-keeper during the game, setting tasks according to the perceived ability of each pair of subjects, and another experimenter made interventions and recorded the children's responses (on paper, audio tape, and occasionally on video). Inter-viewing individuals in this context and requiring them to make the steps of their algorithms explicit, allowed us to gather a significant amount of data on how children calculate mentally. Their actual preferred methods were compared with known common ones and found to be very similar. All of the (correct) informal methods obtained during the fieldwork were specified as rules and entered into the tutor's production system.

The Shopping on Mars activity was tested during all stages of its development on the Archimedes microcomputer and many iterations were required to make the system usable. (Much of the iterative work related to simplifying the interface and improving the system's response time.) The activity proved very successful with children in this age group (8–12). With only one or two exceptions, all of the children tested could (a) quickly and easily learn to use the mouse-driven cursor to drag objects, (b) use the direct manipulation interface to carry out shopping transactions involving manipulation of screen money, (c) operate an on-screen calculator, and (d) understand the notion of containers. The activity was very popular with the children and we believe that the effort spent on good quality graphics was repaid by their positive response.

4.2.2 Description Language

One aim of the fieldwork was to measure the success of our description language as a medium for two-way communication of informal algorithms. We found that the children were usually able to achieve a reasonable grasp of all of the tools and functions of the GADL within a half-hour session. The skills they could demonstrate included (a) forming numerical strings and expressions, (b) using a variety of tools to transform number and (c) operating a scrolling bar enabling them to review an arithmetic procedure. 15 of our subjects were given a second session with GADL after time periods ranging from 1–11 weeks. We observed that after a few minutes of recap by the experimenter, most (9) of the children could recall the interface sufficiently well as to be able to use it again to communicate their informal algorithms to the computer, with minimal prompting from the experimenter. This was true even after the longest time periods.

We have also tried using the GADL to teach children new informal methods for addition, subtraction and multiplication. Those methods were of similar or greater complexity than the ones we had previously observed the children using. They consisted of alternative – usually more efficient – strategies for solving problems similar to or harder than those which the children had successfully solved before. 12 of our previous subjects (aged 9–12) took part in a preliminary study. The experimenter used the GADL and a limited verbal description (of the kind we later represented using a dialogue box and synthesised speech) to demonstrate algorithms for the various operations. Each individual was shown algorithms which we had not previously observed them using. We found that in 94% of cases (n = 34), the children could pick up a new method from the GADL description and could apply it to a new problem. In four of the successful cases, it was first necessary for the experimenter to repeat the new method once.

The results indicate that the GADL has a great deal of potential as a means for communicating sophisticated mental arithmetic algorithms to children. The majority of children relate easily to the metaphor of numbers as manipulable objects and quickly acquire skill at using the interface; direct manipulation works here because of the success of this metaphor. A spin-off from the data resulting from these studies was the specification of a diverse set of common mental methods.

4.2.3 Environment

The design and evaluation of Shopping on Mars are both closely tied to the standard school setting. We also carried out field trials in an open access setting on a University campus during the school holidays where children were invited to use the package if they wished. Fifteen children aged between 7 and 13 took part.

The findings have some similarity to those in the conventional school settings but it was possible to gauge more accurately the children's attitude towards the program.

A large proportion of the subjects had previous computer experience gained either at home or at school. The two most notable assets of the interface which were commented upon by the children were the graphics and the use of the mouse to communicate their intentions to the system. The surprising thing was that most had never used a "mouse" before, but after a short period of use, said that they preferred it to a keyboard. From our observations, children who came to play Shopping on Mars came because they wanted to take part in an adventure, that of shopping on Mars itself. Embedding maths in the game proved a success because it showed children that maths could be fun. One child said "that maths in a game was fun", and another said "It was fun because I didn't realise I was doing maths". Such feedback from the children is very encouraging. Many of the children said that they found maths in school boring because of the way it was presented.

The summer workshop for children attempted to provide an environment where children could be provided with a "window" to an adventure world, such as Shopping on Mars. The children who played the game found it exciting, learnt something new even though they did not all wish to adopt the new methods, and all expressed a desire to play the game again given the chance to do so.

The pupils using Shopping on Mars were a very heterogeneous group with regard to age, ability and the time of day they wished to use the program. Initially it was harder for them to learn to use the package than the pupils in the school field trials because of the lack of support from teachers familiar with the pupils. However, it became clear that the open access setting has many benefits. The children learnt from the older children and from those more experienced with the package. They learnt by asking other children for help, by listening to them and insisting on explaining, and by hovering in the vicinity of the computer and eavesdropping. The two most encouraging findings of the open access field trials are the way that the children taught each other how to use the harder parts of the interface (such as GADL), and the way that their motivation developed and they returned to voluntarily use Shopping on Mars.

5. Conclusion

Before implementing a teaching package it is necessary to understand the variation in learners' competence of taught and informal strategies such as carrying and partitioning but also of the fundamental arithmetic concepts which underpin those strategies such as commutativity and associativity. We have to understand the developmental processes whereby these concepts and strategies are learnt so that we teach appropriately and also so that we can take best advantage of what the learner knows.

Following appropriate longitudinal studies, we build and test computational packages. The two pieces of 'technology' most used in our work are the specification of production rule models for learner competence and the development of direct manipulation interfaces to monitor and minimise the cognitive load on the learners. Then we complete the circle and return to classroom and informal educational settings and test our packages with learners certain that the richness and variation in children's arithmetic strategies will oblige us to change and refine our packages.

References

Baroody, A.J. (1984). The case of Felicia: A young child's strategies for reducing memory demands during mental addition. Cognition and Instruction, 1(1), 109-116.

Baroody, A.J. & Gannon, K.E. (1984) The development of the commutativity principle and economical addition strategies. Cognition and Instruction, 1 (3), 321 - 339.

Evertsz, R., Hennessy, S., Devi, R. (1988). GADL: A Graphical Interface for Mental Arithmetic Algorithms. CITE Report No. 49, The Open University.

Gelman, R. (1977). How young children reason about small numbers. In N. J. Castellan et al. (eds.), Cognitive Theory, vol. 2. Hillsdale, NJ: Erlbaum.

Gelman, R. & Gallistel, C.R. (1978). The Child's Understanding of Number. Cambridge, Mass.: Harvard University Press.

Hennessy, S. (1986). The role of conceptual knowledge in the acquisition of arithmetic algorithms. PhD thesis. University College, London.

Hennessy. S., O'Shea, T., Evertsz, R. and Floyd, A.: 1989. An intelligent tutoring system approach to teaching primary mathematics. Educational Studies in Mathematics, 20, 273-292.

Jones, A., Scanlon, E., O'Shea, T. (eds.) (1987). The Computer Revolution in Education. Harvester Press, Brighton.

O'Shea, T. (1979). Self-Improving Teaching Systems, Birkhäuser, Basel.

O'Shea, T. and Smith, R.B. (1987). Violating the laws of nature: Experiments in understanding physics by exploring alternate realities. Proceedings of CAL87, Glasgow.

Resnick, L.B. (1983). A developmental theory of number understanding. In: H.P. Ginsburg (ed.), The Development of Mathematical Thinking. London: Academic Press.

Young, R. and O'Shea, T.: 1982, Errors in children's subtraction, Cognitive Science 5, 153-177.

Appendix

Details of subjects and results of study on commutativity stages, generalization to subtraction and 3-term addition.

ability is one of high, medium or low (determined by the teacher).

subtraction column: G for generalization, SFL smaller from larger, 0 zero, NP not possible, - for those not tested.

grouping: 1 grouping, 2 explicit knowledge of grouping but did not use it, 3 applied comm. to first two terms, 4 no evidence of transfer, 5 did not know comm. but used COL, 6 left to right strategy (did not know comm.).

	subject	age	ability	comm. stage	subtraction	grouping
1.	AAK	6.6	H	ii	G	6
2.	DR	6.4	H	iii	NP	4
3.	STM	6.3	H	ii	G	6
4.	JMS	6.8	H	i	NP	6
5.	AAN	6.9	H	ii	NP	6
6.	KD	6.8	H	ii	-	-
7.	MC	6.11	H	i /ii	0	-
8.	YPR	6.3	H	iv	-	1
9.	AS	6.10	H	iii	NP	4
10.	AS	6.10	H	iv	NP	1
11.	AK	6.8	H	iv	0	1
12.	AN	6.7	H	iii	G	2
13.	MSN	6.11	M	iii	0	2
14.	VS	6.6	H	iv	G	1
15.	RDP	6.5	H	iv	0	1
16.	AAN	6.10	M	iv	NP	1
17.	RDD	6.5	M	i	-	6
18.	AS	6.3	M	iv	0	1
19.	PR	6.9	M	i	-	6

20.	MS	6.6	L	i	-	-
21.	PS	6.7	L	i	-	-
22.	PA	6.4	L	i	-	-
23.	AD	6.9	M	i	-	-
24.	JS	6.11	H	iv	-	1
25.	SP	6.8	M	iv	0	1
26.	RM	6.6	M	iv	G	1
27.	AK	6.8	M	ii/iii	G	-
28.	AS	5.8	H	i	-	-
29.	M	5.7	M	i	-	-
30.	S	6.6	L	ii	-	-
31.	AK	5.9	L	i	-	-
32.	K	6.8	L	ii	-	-
33.	WA	6.9	H	ii	0	-
34.	PA	6.3	H	ii	0	-
35.	B	5.9	M	i	-	-
36.	APS	7.6	H	iv	G	1
37.	AS	7.6	H	iv	SFL	1
38.	NV	7.8	M	iii	NP	4
39.	PKP	7.5	L	iii	G	2,3
40.	RD	7.8	L	iii	G	-
41.	SR	7.8	H	i/ii	np	6
42.	NPS	7.10	H	iv	NP	1
43.	HN	7.6	L	iv	-	1
44.	PP	7.5	M	i	0	6
45.	SL	7.1	H	ii	NP	6
46.	RL	7.2	M	iv	G	1
47.	AS	7.3	H	iv	NP	1
48.	MS	7.1	H	iv	NP	1
49.	AC	7.2	H	iv	NP	1
50.	TAG	7.3	M	iii	NP	3
51.	APL	7.0	M	iv	NP	1
52.	SL	7.10	H	iv	SFL	1
53.	SS	7.5	H	iv	G	1
54.	SH	7.6	H	iv	SFL	1
55.	RK	7.9	L	iv	-	1
56.	NR	7.2	M	i	-	6
57.	AA	7.9	H	iv	-	1
58.	SR	7.9	L	iv	G	1
59.	RP	7.8	L	iv	-	1
60.	SR	7.8	L	iii	NP	3
61.	VK	7.1	L	i	-	-
62.	KL	7.2	L	i	-	-
63.	AA	7.6	M	iv	0	1
64.	RK	7.3	M	i	-	-
65.	JR	7.0	M	ii	-	-
66.	AK	7.0	L	i	-	-
67.	R	7.7	H	iv	-	1
68.	ML	7.5	H	ii	0	-
69.	NNK	8.3	M	iii	G	1

70.	AS	8.0	M	ii	0	-
71.	NTR	8.1	M	ii	-	-
72.	RP	8.6	H	iii	NP	3
73.	RP	8.6	H	iii	NP	3
74.	AK	8.0	H	iv	NP	1
75.	IRL	8.5	L	i/ii	0	6
76.	SD	8.3	M	i/ii	0	6
77.	AS	8.6	H	iv	NP	1
78.	PD	8.4	M	ii	-	5
79.	SK	8.7	M	ii	G	1
80.	M	8.8	H	ii	0	6
81.	KN	8.7	H	iii	-	1
82.	NS	8.4	H	ii	-	6
83.	AS	8.0	H	iv	0	1
84.	RK	8.2	M	iv	G	1
85.	GC	8.0	L	iv	-	1
86.	SKL	8.2	M	iv	-	1
87.	JN	8.4	H	iv	SFL	1
88.	RP	8.0	H	iv	NP	1
89.	AB	8.6	M	iv	NP	1
90.	ST	8.6	H	iii	0	-
91.	AC	9.2	L	ii	-	-
92.	RP	9.5	M	i/ii	-	6
93.	SSD	9.3	M	ii	NP	-
94.	LD	9.0	M	ii	0	3
95.	R	9.11	H	iv	-	1
96.	SM	9.10	H	iii	0	3
97.	N	9.9	H	iv	SFL	1
98.	NC	9.0	M	iv	0	1
99.	SP	10.7	H	iv	-	1
100.	F	10.2	H	iv	-	1
101.	RK	10.10	H	iv	-	1
102.	M	10.11	H	iv	-	1
103.	JH	10.2	H	iv	NP	1
104.	AD	11.0	H	iv	-	1
105.	N	11.4	H	iv	-	1

Cognitive Interpretation of Microworld Operations

Tommy Dreyfus[1] and Baruch Schwarz[2]

[1] Center for Technological Education, P.O. Box 305, Holon 58102, Israel
E-mail: tommy@barley.cteh.ac.il

[2] School of Education, Hebrew University, Mount Scopus, Jerusalem, Israel
E-mail: msschwar@pluto.mscc.huji.ac.il

Introduction

Learning environments, or mathematical microworlds, have been claimed to be the prime choice for supporting those learning processes which are aimed at understanding the properties of mathematical objects and the relationships between them which are so important in mathematics. The researcher can investigate such claims by observing the students understanding before and after the learning experience and draw general conclusions from his observations about the overall influence of the microworld. Such investigations may confirm the claim made above, but they will not promote our insight into the mechanisms through which the students acquire their understanding. These mechanisms are linked to cognitive processing on a much more fine-grained level. One way to investigate them, is to establish a more detailed connection between microworld operations and cognitive phenomena.

In this paper, we will refer to a case study in which it was possible to build such a detailed connection between the cognitive and the computational (Schwarz and Dreyfus, 1993). This study dealt with a multiple linked representations learning environment for functions. A model was constructed to measure students transfer of information between representations during the process of solution of one specific problem. The indices in this model were given a cognitive interpretation in terms of the role representations play in the students concept of function. This cognitive interpretation was successfully validated by comparing the model's results with the interpretation of experts who did not know the model.

The work in this case study was specifically geared to a particular learning environment and a particular concept. The question arises what aspects and methods of this work can be generalized to other environments and to other concepts. One way to attempt to answer this question is to try to build a similar model within the framework of another environment, possibly one with a different character. We propose to do this for a learning environment of constructions and transformations in three-dimensional geometry. Although work in this direction is still in a preliminary stage, it will be presented here because of its relevance for student models in geometry.This will also enable us to compare the two cases and thus to speculate what features of the model construction can be generalized and what conditions need to be satisfied in order to make such a model feasible.

Background

Following Thompson (1987), a microworld is a system composed of objects, relationships among objects, and operations that transform the objects and relationships. In such an environment, students explore their understanding of the subject matter in a manner similar to that by which scientists test their conjectures about the way the world works. A direct consequence is that microworlds do not teach by themselves and they do not provide instruction. Another feature of such microworlds is that they are operational, i.e. that any action of the student in the microworld is organized in operations on the objects of the microworld. This will be important later.

It is not easy to see microworlds with such a definition as part of a curriculum. If the system does not teach, how does the student learn? Therefore, in addition to the computerized environment which encourages exploring, conjecturing, etc., the present concept of microworld includes a sequence of structured activities designed to help the learner develop the concept(s) under consideration. We adopt this definition here.

Thompson (1987) also framed a program for the design of computerized environments. Figure 1 shows the four principal components of his program.

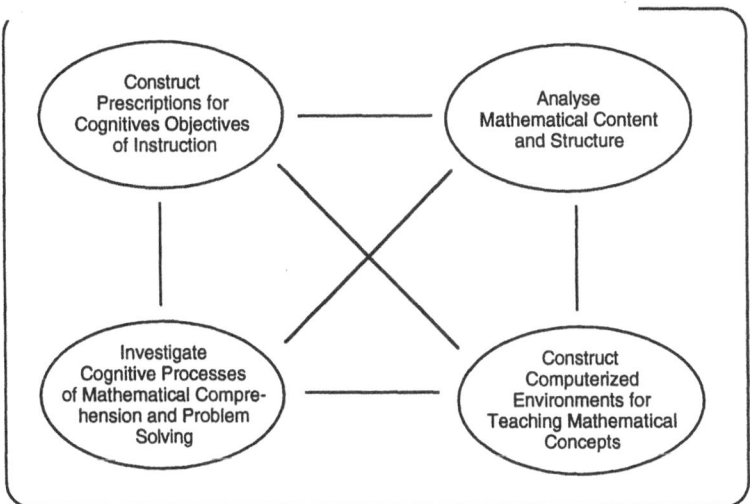

Figure 1. The four components of a research program on understanding and improving mathematics teaching and learning (Thompson)

Such a program demands a considerable effort in several respects:
- constructing a pilot microworld, taking into account the mathematical structure of the concept(s) and cognitive learning objectives;
- gathering data from observation of students working with the microworld;
- building a model of student learning;
- modifying the pilot microworld

This effort pays off by giving us the possibility to look into the connections between cognitive and computational aspects. Indeed, since the technical work will to a large extent be carried out by the computer, and the microworld operations have been constructed in precise correspondence to the cognitive objectives, we may hope to be in an ideal situation to infer from the student's use of these operations conclusions about his or her cognitive state. In other words, since the students concern when using the microworld, centers directly and rather exclusively on the concept(s) dealt with in this microworld, his or her understanding of these concept(s) can be accessed and assessed relatively easily.

The tendency, today, is to develop microworlds which have an "intelligent component". (These are not usually tutoring systems.) The basic idea of the intelligent component is to help students choose strategies for problem solving and to give them feedback based on their present state. The meaning of "present state" has to be defined precisely, by a model. We will not enter here any of the deep and difficult questions which concern the construction of such models. We want to relate only to one particular aspect of such models, namely, that they will not be educationally efficient unless they are based on a sound cognitive foundation. It would be naive to think that a student model could be based only on an analysis of the mathematical topic to be learnt. It must be based on detailed experimental evidence about students' processes of understanding and difficulties when learning the topic.

Once this is achieved, and a student model exists, the question must be asked how the student's state, his or her description in terms of the cognitive model, can be identified on the basis of his or her interaction with the computerized learning environment. It is this last question which we want to address in this paper. Since the environments to be discussed do not have computerized student models, we slightly modify the question: Suppose we have a learning environment as well as cognitive information about students interacting with this environment. To what extent, and how, can we use a student's interaction with the environment, i.e., the sequence of operations he or she carries out (or, more mechanically expressed, the sequence of key strokes,) to draw conclusions about his or her cognitive state?

In fact, even if the design of the microworld operations were perfect, one could not expect a clean correspondence between cognitive state and microworld operations. Problems of inconsistency on the part of the students, of noise, and of cognitive style arise, and every trace of a learning process comprises these components. Even if protocol analysis of tasks can diagnose some of them, their disturbing effects must be taken into account when trying to see, through the sequences of microworld operations, the cognitive states in the learning process of the student. This difficulty is enhanced when one wants to see the ongoing process of change of such cognitive states. What we claim is that in spite of these difficulties there exist some conditions of applicability which must be satisfied if one wants to establish a correspondence between microworld operations and cognitive states. These conditions concern the type of microworld, the type of task, the design of the recording and even the type of student.

We will base our argument on the comparison between two microworlds on two different mathematical topics and with different types of tasks.

Case 1: Transfer Between Function Representations

Only a brief account of this work is given here; it has been described in more detail elsewhere (Schwarz & Dreyfus, 1993).

A multiple linked representations microworld for functions, named TRM (the Triple Representation Model), was used for this investigation. Three representations were implemented: a tabular, a graphic and an algorithmic one. The use of the microworld was integrated in a ninth grade curriculum. After an instructional period of about four months, students were given a standard open box problem (find the dimensions of the largest open box that can be made from a given piece of cardboard). All microworld operations they carried out during the solution of this problem were recorded and processed. An index was constructed which measured whether and how well students progressed on their solution path; it was called the Quality Index. Another index referred specifically to the stages at which students passed from working in one representation to working in another one; this index, called the Transfer Index, measured whether, when working in a particular representation, students used information they had collected previously while working in other representations. In other words, the Transfer Index measured the transfer of information between representations. These indices were given a cognitive meaning in terms of the role representations play in the students' concept of function: on the basis of the values of a student's indices, we interpreted whether for this student the representations were signifiers for the function, whether they were themselves the signified objects or not even this. Representations are signifiers for the function if the student perceives general properties of the function through the representations (and, consequently, can use these properties in other representations); if, on the other hand, a student is able to understand properties of the function within a representation, without being able to use them in another representation, the representations are themselves the signified objects. This cognitive interpretation was successfully validated by comparing the model's results with the interpretation of experts who did not know the model.

The computation of the indices in the model was based, fairly directly, on the choices students made of the intervals within which to look for the solution.

For example, consider a student who, at a certain stage of the solution process, wants the computer to draw a graph of the volume of the box as a function of the size of the square corner being cut off from the cardboard. When asking for the graph, the student needs to decide on and specify an interval within which the graph is to be drawn. There is, at any stage of the solution process, an "ideal" interval that could be chosen, given all the information collected previously. There may be several reasons for the student to choose an interval different from the ideal one; for instance, (s)he could round of the values of the end-points of the ideal interval; or (s)he could decide to search in a smaller interval, based on sound intuition that the solution is in this smaller interval anyway; or (s)he could choose a larger interval to get a more global picture of the behavior of the function; or (s)he could choose a much larger interval because (s)he does not take previously collected information into account; or (s)he could choose a much smaller interval, because of misinterpretation of previously collected information; this list could be continued at will. In building the model, we had to assign values to the indices which were based on the chosen interval alone, rather than on the reasons which a student may have had to choose a particular interval. Some delicate decisions had thus to be made about which intervals should be considered "good" choices, i.e., choices which took previously collected information correctly into account (contribution to the Transfer Index) and led to progress toward the solution (contribution to the Quality Index). There is necessarily some arbitrariness in these decisions. Therefore we made several different "reasonable" decisions and checked the resulting differences in the Transfer and Quality Indices assigned to the students as well as in the classification concerning the role representations play in the students concept of function. Although quite a lot of differences resulted in the local components from which the indices were computed, the values of only three indices changed, out of the 86 we computed in the study (two indices for each of 43 students), and none of the cognitive classifications changed.

Differently stated, we found that the model we had constructed was stable under small perturbations of a certain kind. Similar stability has been found to hold with respect to other perturbations such as pressing a key inadvertently, or to the influence of the students cognitive style.

It thus turned out that the global aspects of the model are stable, even though the local ones need not be. But since all interpretations were based on these global aspects, local instability was irrelevant.

For obvious reasons, it is very important that the cognitive interpretations of a model are not sensitive to small perturbations in the data and thus to the kind of arbitrary decisions that have been exemplified above. In the case of the model for transfer, this stability has been achieved by means of the fact that the Transfer and Quality Indices were computed on the basis of many microworld operations, typically about twenty. As a consequence, local changes had little effect, and often even averaged out. In cognitive terms, the computation of the Indices from many operations (which were carried out by the student sequentially), means that the units of student activity that were interpreted were sets or sequences of operations rather than single operations. Taking a set of operations as the unit of interpretation is especially reasonable and easy to carry out if the students' activity naturally decomposes into stages or episodes. In the model for transfer this was the case: The work carried out by the student in one representation was taken as an episode; a new episode begins when the student changes the representation in which he or she works.

Case 2: Stereometry

Solving stereometry problems requires thinking in two representations, visual and analytic; the combination and integration of these is not easy and needs serious attention in the teaching process. Although specific learning difficulties in stereometry have not been investigated, there is some research work on the understanding of the transition between three-dimensional objects and their two dimensional representations; this transition is bi-directional: imagine, and draw projections of a given solid and interpret a plane figure as the projection of a three-dimensional solid. Much of the research on this transition dealt with polycubal solids and students' ability to create and interpret two-dimensional representations for them; in most cases, it was found that students had great difficulty in successfully communicating visual information as well as in interpreting two-dimensional drawings of polycubal solids (for a review see Sect. III of Hershkowitz, 1990).

Parzysz (1988), on the other hand, investigated students' implicit rules used in the transition between two and three dimensions in the case of a square base pyramid. He reported that his students tended to confound the three-dimensional figure drawn with the two-dimensional one having the same representation. In drawing, the main difficulties he found originated from a conflict between what is seen and what is known; for instance, the base of the pyramid was more often rendered as a square or diamond than as a rectangle or parallelogram. Osta (1987) used the dynamic potential of the computer to teach spatial topics, specifically polycubal solids, to grade 8 and 9 students and to observe their progress. She found that, in the course of the instructional sequence, they passed from using more perceptual criteria to using more geometrical ones.

In order to aid students in this transition form the perceptual to the geometrical and to avoid the conflict between what is seen and what is known, a microworld for stereometry, named Stereometrix, has been designed (for a detailed description see Dreyfus & Hadas, 1991). Stereometrix has two main types of operations: Constructions and Transformations. Constructions are operations that change a given solid, e.g., perpendiculars from points to either a line or a plane. Transformations are operations that do not change the solid but only its position in space and thus its representation on the screen, e.g., rotations. A special type of rotation was introduced which makes it possible to rotate the solid in such a way that a given plane becomes the screen plane. This special rotation has turned out to be one of the most powerful features of the microworld; it can be used to check whether a particular point or line lies in a plane, whether two given lines intersect, etc. An additional feature of Stereometrix is that it allows the student to save the sequence of operations he or she used in a construction and to replay it either on the same or on a different solid. The construction is thus decoupled from the object on which it is carried out. This feature is useful in order to find what remains invariant and what changes as one passes from a special case to a more general one.

Most typical activities with stereometrix are of the type: Predict and check. The learning experience is then based on the conflict arising if the prediction was incorrect.

Such activities are not suitable as a basis for cognitive interpretation because they are usually solved in a small number of microworld operations, each of which may follow from and thus express a considerable amount of thinking effort, i.e. of cognitive information. In such a case, the translation of cognitive activity into microworld operations is irreversible.

There are two types of exceptions to this. One is a task that could possibly be used for the construction of a computational model for the transition between two- and three-dimensional representations: Given a solid, represented on the screen in a certain projection, turn it so that another projection, the goal projection results. This task seems similar to the maximum problem used in TRM in that it provides for a gradual approach to the solution; indeed, it is not difficult to find a suitable measure for the distance from the goal, and thus find an analog to the Quality Index. In spite of this, we have not been successful in finding a good correlation between the microworld operations and the cognitive skills under discussion. This can be understood as follows: In the open box task, the student is required to find the answer with an accuracy of 10^{-4}. In the TRM microworld, this can be achieved only by turning, at the end of the solution process, to the algebraic representation. The student is thus forced to change representations at least once (most students in fact changed three or four times). This change of representation is directly linked to the cognitive question under investigation. In other words, the task is such that an operational process leading to its solution can be translated directly into cognitive terms. Such a direct link could not be established for the stereometry task. Moreover, in the stereometry task, there is no clear structuring of the students' sequence of operations into episodes, as there was for the TRM task. We stress again that these episodes were essential in constructing the cognitive interpretation.

A second type of stereometry task held some promise for being suitable for cognitive interpretation of the operational solution process; this type is constructive, for example: Given a right prism ABCDEF with a known isosceles triangle (AB=AC) as base, and given the angle between the face diagonal AE and the face ACFD, find a procedure to compute the volume of the prism (Fig. 2). (The actual computation of the volume is lengthy and irrelevant here.) The skills needed for solving this problem include the construction of additional elements, the point G, and the identification of suitable planes, such as AEG, in which computable triangles appear. These elements need to be visualized and constructed and/or projected onto the computer screen with Stereometrix. Choosing, visualizing and constructing these elements are central cognitive activities for problem solving in stereometry; but they find their expression in the solution process as single microworld operations, at best. Consequently, the entire problem solution is likely to contain only a few significant operations. Although a cognitive model could possibly be built on this basis, it would be hopelessly unstable under perturbations.

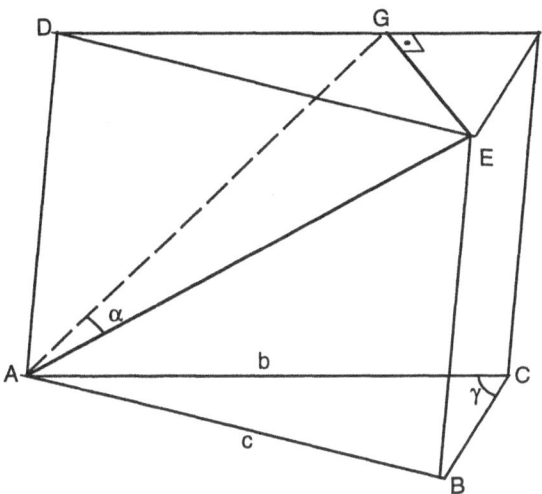

Figure 2. Given b = c, α, γ, how can you compute the volume of the prism?

Conditions for Interpretability of Microworld Operations

On the basis of the two case studies described above, we will now discuss conditions that appear to be necessary for a cognitive interpretation of microworld operations. Whether these conditions are also sufficient, will be left open.

In some cases, not enough information can be gleaned from students' work. In other words, solving some tasks in some microworlds yields sequences of microworld operations which are too short for the construction of a cognitive model. This may be due either to the microworld or to the task (or both). An example is the construction type task in Stereometrix. But it is easy to find tasks which are insufficient in this respect also in TRM. For instance many function transformation tasks (such as shifts, stretches or reflections of graphs) are of the predict and check type, and we have seen that this type of task is unsuitable for interpretation. In other words, the microworld *and* the task must be such that the students' progress necessitates many operations (order of magnitude: 20 to 50). Only in such a case do we obtain a sufficiently fine mesh size of information to catch the "cognitive fish". Operations must not be sparse in the sense that a lot of student thinking intervenes between successive operations. Thus we postulate Abundance of operations as a first condition.

A second condition concerns the kind of information provided by the sequence of operations; In TRM, the sequence of operations *is* the solution; in Stereo-metrix, it allows you to see the solution. When seeing it, you may not need to carry it out on the computer, and therefore the sequence of microworld operations may not contain the interesting information, the one that can be interpreted in cognitive terms. But again, not the microworld alone is important here. The choice of the task is of prime importance. We need to design problems which are suitable from the point of view of cognitive research, i.e. which address interesting cognitive questions and which also give the possibility of constructing a computational model. That is, the tasks need to exploit the features of the microworld forcing the students to detail their knowledge. We may want the task to dictate some operations. This is difficult, and needs to be done by means of cleverly devising the task. Possibly, such tasks exist only in some microworlds and not in others. The requirement for accuracy of 10^{-4} described above is of exactly this kind: It forced transition from one representation to another one. We thus have a second condition: Type of task and ensuing Quality of information.

A question of control is associated with this condition of Quality. Microworlds usually give the student full control over his or her actions. In order to investigate a particular aspect, such as transfer, we would like to force the students into making certain steps (e.g., switching representations). On the other hand, we would also like them to be in full control of their work with the microworld. In other words, we want to remote control them without influencing their performance. Again, both the microworld and the task must be designed to achieve this. As an example, consider graph sketching in TRM. When a student requests to draw a graph, (s)he needs to specify the domain, in which the graph is to be drawn. Obviously, this could be at least partially automated, but it was a conscious design decision of the microworld, not to do this for didactic reasons. The cognitive researcher profits from this design decision, because the students have to explicitly detail their knowledge. In fact, the cognitive model for transfer described above specifically depended on the fact that students specified domains.

Last but not least there is a question of structure. The question is related to the identification of a unit of interpretation in the sequence of microworld operations. In other words: Can the students' progress on the solution path be structured into separate stages, which are interpretable? It has been shown above that an essential ingredient for the interpretation of the transfer sequences was that they were well structured into episodes. Condition three is thus: Structure.

In summary, three conditions have been identified that are necessary in order to build a cognitive model from sequences of microworld operations: Abundance, Quality, and Structure. We have seen above that Abundance is also a prerequisite for the stability of the cognitive model: A stable model gives globally correct cognitive information about the student, and this information is independent of minor changes in the sequence of operations. A stable model can provide the cognitive information which enables judgement of the students present state of knowledge; it may thus eventually lead to the construction of a microworld able to react to student actions in an appropriate way.

Conclusion

In this paper, we have disregarded the ideological question whether or not it is worthwhile "to model the student". This question is addressed in other chapters in this volume (Hoyles, Schwartz). Instead, we have posited that modelling the student is worthwhile, and we have investigated under what conditions this undertaking has a chance of success.

We have identified three necessary conditions for building a computational model in the sense described above, built on student progress during problem solution as expressed by the actions of the student within a microworld. The three conditions are Abundance, Quality and Structure. We believe that if these three conditions are satisfied, there is a good chance that we can find, in a typical students problem solution, sequences of microworld operations which can be given a meaningful interpretation in cognitive terms.

We have discussed one case study, where a connection between the cognitive and the computational aspects of students work with a microworld has been established to a considerable depth. And we hope thus having contributed to the closure of the gap between the principles of student modelling and existing mathematical microworlds which has been identified at the Grenoble workshop.

References

Dreyfus, Tommy, & Nurit Hadus (1991). Stereometrix – A learning tool for spatial geometry. In: W. Zimmermann & S. Cunningham (eds.) Visualization in Teaching and Learning Mathematics (pp. 87-94). Washington, DC: MAA Notes Vol. 19.

Hershkowitz, Rina (1990). Psychological aspects of learning geometry. In: P. Nesher & J. Kilpatrick (eds.), Mathematics and Cognition: A Research Synthesis by the International Group for the Psychology of Mathematics Education (pp. 70-95). ICMI Study Series. Cambridge, UK: Cambridge University Press.

Hoyles, Celia (1996). This volume.

Osta, Iman (1987). L'outil informatique et l'enseignement de la geometrie dans l'espace. In Jacques C. Bergeron, Nicholas Herscovics & Carolyn Kieran (Eds.), Proceedings of the Eleventh International Conference for the Psychology of Mathematics Education, Vol. II (pp. 31-38). Montreal, Canada: Université.

Parzysz, Bernard (1988). "Knowing" vs "Seeing". Problems of the Plane Representation of Space Geometry Figures. Educational Studies in Mathematics 19(1), 79-92.

Schwartz, Judah (1996). This volume.

Schwarz, Baruch, & Tommy Dreyfus (1993). Measuring integration of information in multirepresentational software. Interactive Learning Environments 3(3), 177-198.

Thompson Patrick W. (1987). Mathematical Microworlds and Intelligent Computer Assisted Instruction. In: G. Kearsley (ed.) Artificial Intelligence and Instruction: Applications and Methods. New York, NY: Addison Wesley.

Calculus Revisited

Maurizio Falcone

Dipartimento di Matematica, Università di Roma "La Sapienza"
Piazzale A. Moro, 2, I-00185 Rome, Italy
E-mail: falcone@axcasp.caspur.it

1. Introduction

In the last decade there has been a big debate on the role of computers in mathematics mainly motivated by the technological revolution that allowed to develop numerical experiments and simulations using even very small computers. Infact personal computers have given to all students and teachers great opportunities (reserved before only to experts in computing and computer science) becoming an essential tool for teaching and research. This new and exciting situation has stimulated many experiments in different directions to understand the impact of new technologies in mathematics but has also generated some negative reactions. The supporters of an extensive use of computers in mathematics maintain that by means of computers it is possible to visualize complex concepts and phenomena receiving interesting hints for the theoretical solution of difficult problems, moreover they assert that the application of numerical methods can give "real" solutions instead of pure existence and unicity results and that computers make possible to present in a classroom a variety of examples and situations improving our teaching. This position has been expressed in many articles and reports on experiences which are still going on, more recently the discussion on the increased interaction between computers and mathematics left the circle of experts to involve the whole mathematical community (see, e.g., [13] and the Jon Barwise column in the Notices of the American Mathematical Society).

On the other hand, there are still some strong opponents who claim that since nobody would be able to prove a new mathematical result by means of a computer these are completely useless for research. Even the visualization of mathematical phenomena, they say, can be misleading: due to numerical errors, computers cannot handle subtle mathematical concepts. Some of them disregard applications since they feel pure mathematicians (see [14] and, for an historical reference to this point of view, [10]).

As far as teaching is concerned probably the best way to analyze the impact of computers on students learning is to make experiments. Infact a number of projects have been carried out in many universities in order to define a methodology for the application of computers in a classroom and in order to develop packages to this end. The experience in Rome has been focussed mainly on the courses of Calculus and Advanced Calculus (respectively at the first and second year of the curriculum) working in strict connection with other european universities and in particular with Paris Sud-Orsay and Leeds within a project which has been supported by the Commission for Education of the E.E.C. starting in 1983 ([6], [7], [8], [12], for reports on other Italian experiences see [1] and [2]). One of the peculiarities of this experiment has been the introduction of programming sessions where students are involved in an *active* programming work. Infact we were (and we are still) convinced that all students attending a scientific degree should be able to use a computer for scientific investigations and should have an informatic background. This goal can be obtained either introducing new compulsory courses or changing the contents of the existing ones. The quite rigid organization of curricula in the Italian universities made easier to follow the second way. As we shall see in the following sections, one of the more interesting aspects of this experiment has been the tentative to modify contents and organization of these courses trying to design a new teaching itinerary.

2. The Contents of the Experiment

As everyone knows, Calculus and Advanced Calculus are foundamental courses whose main objective is to develop logic-deductive capabilities presenting mathematical results and techniques necessary in the next courses. The organization of these courses provides general lectures and practical exercise sessions. The introduction of computers entailed some modifications in both of them since computers have been used to *present and visualize mathematical phenomena* as well as to *solve problems* implementing numerical methods.

We developed demostration programs to illustrate various aspects of mathematics taking advantage of the graphic facilities available on personal computers. The possibility of seeing mathematical objects on a screen greatly increases *the role of geometry* in the learning process also for those courses which seems to be quite far from it. Computers can give you the chance to reverse the traditional definition-examples sequence allowing you to achieve the right mathematical definition *after* having introduced a concept by means of many concrete examples. The software for demonstrations has been the main tool to build a real *Laboratory of Mathematics* in which the teacher makes different experiments discussing their mathematical meaning. Functions, sequences, solution trajectories of ordinary differential equations, curves and surfaces can be exhibited to students showing them the wide variety of behaviours and situations hidden in a mathematical definition. Let us give some examples.

Among the properties of real functions of one real variable the regularity properties play a very important role. First of all students must understand what a function is and accept that a function can be defined in different ways on different intervals retaining as well some regularity properties. The function

$$f(x) = \begin{cases} \exp(x) & \text{if } x < 0 \\ (x-1)^2 + 1 & \text{if } x \geq 0. \end{cases}$$

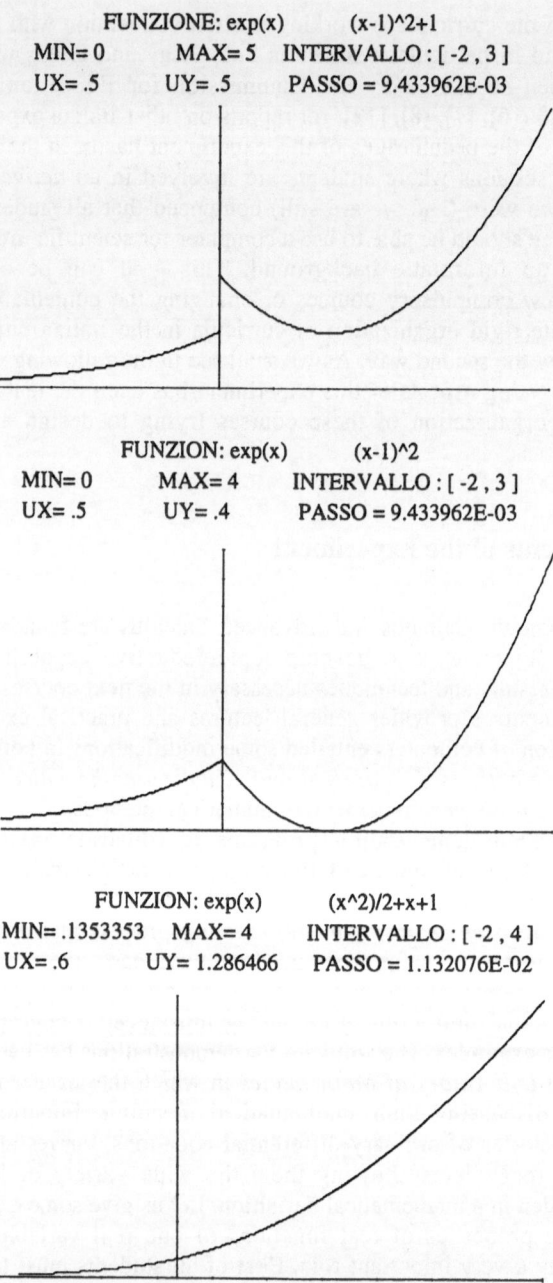

Fig. 2.1 Functions with different regularity properties

X(S)= 2*sin(sqr(2)*s)   Y(S)= 2*sin(s+3)
PARAMETRO IN (O , 60)   PASSO : .1
UX = .7233659   UY = .3999968

X(S)= 2*sin(sqr(2)*s)   Y(S)= 2*sin(s+3)
PARAMETRO IN (O , 160)   PASSO : .1
UX = .7233681   UY = .399998

X(S)= 2*sin(sqr(2)*s)   Y(S)= 2*sin(s+3)
PARAMETRO IN (O , 300)   PASSO : .1
UX = .7233681   UY = .3999981

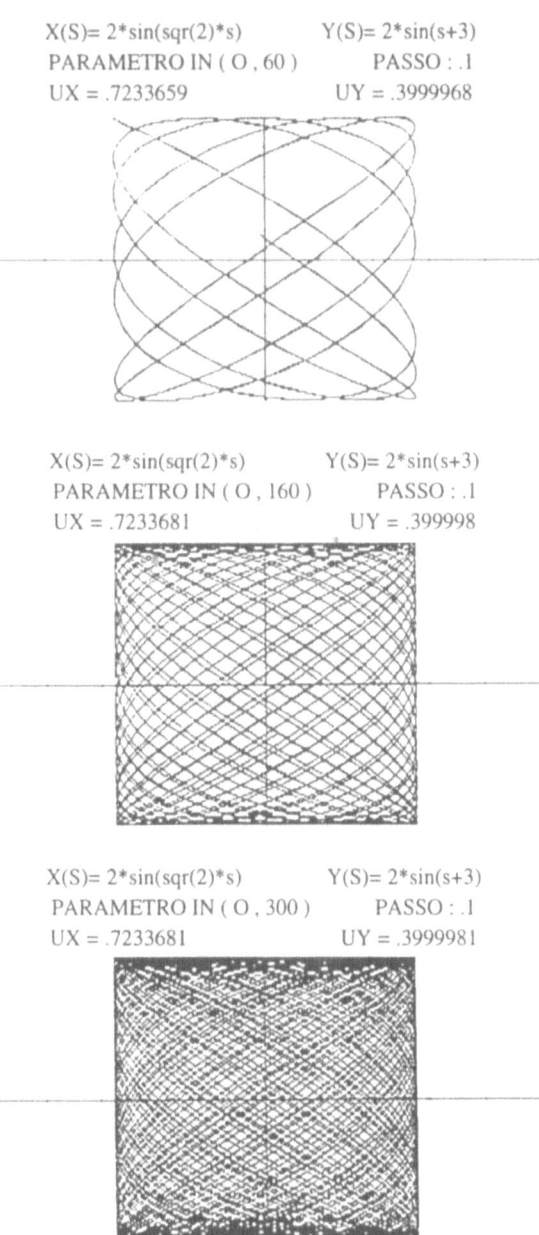

Fig. 2.2 The Lissajou curve filling the square

is defined everywhere on **R** but is discontinuous at the origin as the first drawing of Fig. 2.1 shows. A slight modification of the previous function, namely

$$f(x)=\begin{cases}\exp(x) & \text{if } x<0 \\ (x-1)^2 & \text{if } x\geq0.\end{cases}$$

leads to a function which is continuous on **R** but not differentiable at $x = 0$. Another modification of the parabola leads to a function differentiable on **R**,

$$f(x)=\begin{cases}\exp(x) & \text{if } x<0 \\ \dfrac{x^2}{2}+x+1 & \text{if } x\geq0.\end{cases}$$

The actual computation of the limits

$$\lim_{x\to0^+} f(x), \quad \lim_{x\to0^-} f(x), \quad \lim_{x\to0^+} f'(x), \quad \lim_{x\to0^-} f'(x),$$

gives to the students the analytical definition of what regularity is while the drawings give them the geometric representation of the same definition.

One can use similar programs to show piecewise C^1 curves and C^∞ curves or turn to more interesting examples. The parametric curve

$$\psi:\begin{cases}2\sin\sqrt{2}s \\ 2\sin(s+3)\end{cases}$$

is obviously a C^∞ curve contained in the square $Q \equiv [-2, 2] \times [-2, 2]$, what is less obvious from the definition is that this curve fills Q when $s \in [0, +\infty]$ (see Fig. 2). This curve represents the motion of a point governed by two coupled armonic oscillators acting on two orthogonal axes and having different frequencies and phases. The motions on the x and y axes can be written separately in the general form

$$x(s) = R_1 \sin \omega_1 s$$
$$y(s) = R_2 \sin(\omega_2 s + \varphi)$$

and it can be proved that the corresponding curve in the plane, usually indicated as the Lissajous curve, is closed whenever ω_1/ω_2 is a rational number and fills the rectangle $[-R_1, R_1] \times [-R_2, R_2]$ otherwise.

The visualization of solution trajectories of systems of ordinary differential equations is one of the most impressive result you can obtain using computers. To explain what ω-limit cycles are one can take the Van der Pol equation

$$\begin{cases}\dot{x} = y - (x^3 + x) \\ \dot{y} = -x\end{cases}$$

with a given initial point (x_0, y_0).

Integrating that Cauchy problem by a numerical method you can see the trajectories while they converge to a periodic orbit (the ω-limit) circling clockwise in the phase plane. The same behaviour appear taking (x_0, y_0) inside

(Fig. 2.3) or outside (Fig. 2.4) the region entoured by the ω-limit. Looking at the plots of $x(t)$ and $y(t)$ you can see the trajectory becoming periodic.

Example demonstration is only one of computer applications. Computers can also be very useful teaching rather abstract concepts, e.g., you can visualize the role of ε and δ in the definition of the limit of a function or in that of uniform convergence for a sequence of functions. Showing to students a series of examples helps them to absorb concepts that can be quite difficult in their abstract mathematical formulation, particularly in the first two years of their curriculum. From this point of view the use of computers is very practical and effective since, if the software used for demostration allow an interactive modification of data, one can easily modify the parameters in a given example showing immediately the changes due to the new choice. Programs of this kind have been used in the classroom by means of a personal computer connected to a tricromic projector or to a liquid crystal display.

Students have also used these programs as a *dynamic blackboard* to make their own investigations, verifying their knowledge and the validity of their intuition. This is very important since effective teaching requires that students be engaged by what they learn. At the very beginning it is crucial to guide these investigations: answers given by computers are not always correct particularly in mathematics and in this respect those who are suspicious on the results obtained by using computers are completely right. Nevertheless, even wrong answers can be useful to motivate a presentation of some topics related to computer applications in science, e.g., the binary and hexadecimal representation of numbers, the axiomatic of real numbers and the floating point algebra, various interpolation techniques. This leads to the second major change which has been experimented.

In the courses of Calculus and Advanced Calculus the accent is usually on definitions and on *qualitative* results which can be proved starting from definitions: convergence, existence, regularity, uniqueness of solutions and so on. Once you show an example to the classroom using a computer they will ask you "What is hidden *in* the program?" and demand informations about the numerical methods that make the program run properly. So you have to look at old problems with a constructive mind presenting new topics related to *quantitative* results. This point is crucial not only for applications but also from a theoretical point of view: simulations on a computer can give important hints in the study of many mathematical models as it happens in modern research in mathematics.

In opposition to the main stream of teaching in Calculus courses, we want to stress that there is no reason to emphasize *only* qualitative results. For example, why focus attention only on methods which lead to an explicit algebraic solution of an integral if, in the great majority of practical applications, these methods cannot be used to compute integrals? Why not give to students some elementary tools to compute a definite integral? Why not ask them to implement simple numerical methods and test them on a number of examples? The topics which have been added to the courses (see table above), though elementary, are sufficient to face typical problems which arise during the courses. This idea has been developed in some textbooks which contain many of those topics (see, e.g., [11] and [5]).

CAMPO X: y-x^3+x   CAMPO Y: -x
Eulero modificato P.I. : (2 , 2) Passo .1 Iterazioni:130
UX : .597395   UY = .3262609

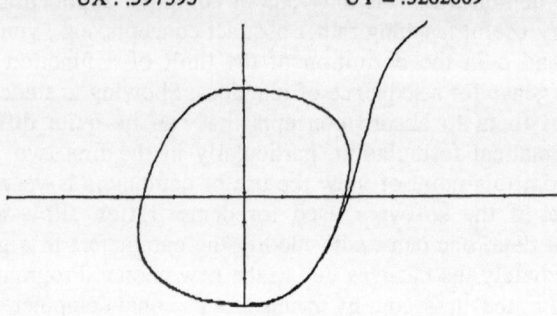

PIANO X/T
UX = 2.3 UY= .3160016 tempo in (0 , 23)

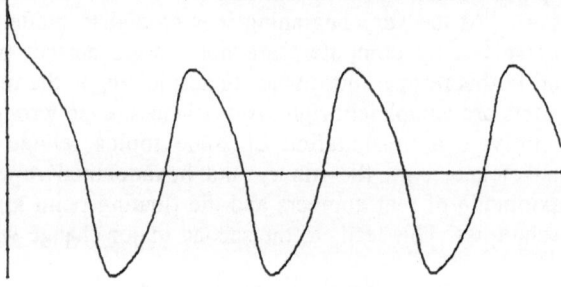

PIANO Y/T
UX = 2.3 UY= .3262609   tempo in (0 , 23)

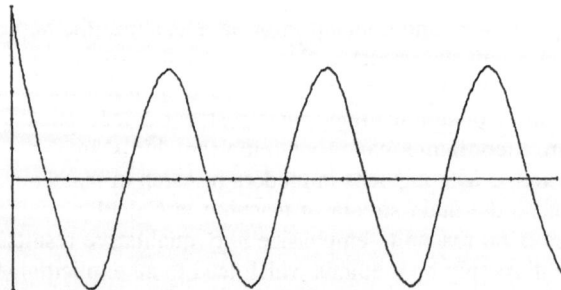

Fig. 2.3 Van der Pol equation: solution starting outside the ω-limit

CAMPO X: y-x^3+x   CAMPO Y: -x
Eulero modificato P.I. : (.1 , .1) Passo .1 Iterazioni: 230
UX : .597395   UY = .3262609

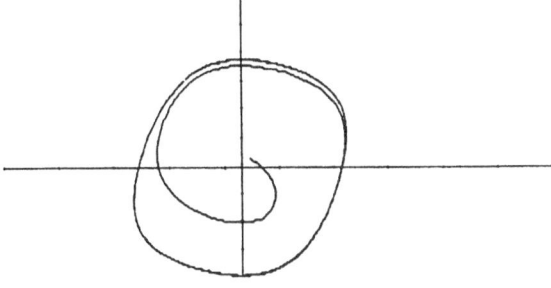

PIANO X/T
UX = 2.3 UY= .3160016   tempo in (0 , 23)

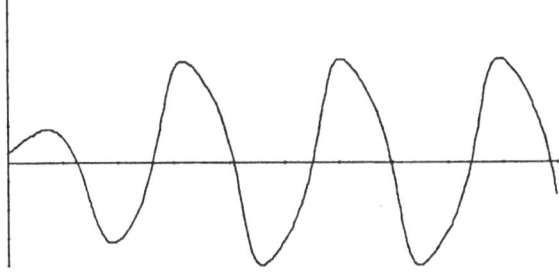

PIANO Y/T
UX = 2.3 UY= .3262609   tempo in (0 , 23)

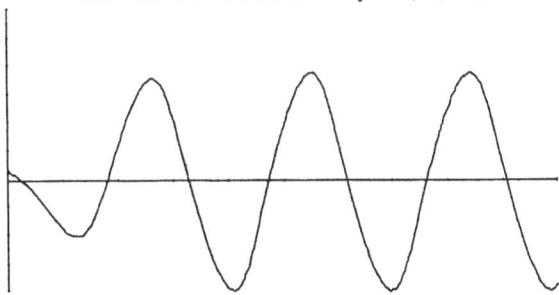

Fig. 2.4 Van der Pol equation: solution starting inside the ω-limit

Students' response to this experiment has been enthusiastic though the number of hours added is relevant if compared with the total duration of the course, which is of about 120 hours. The first 6 hours were devoted to an introduction to programming and to a very rapid presentation of the BASIC language, at this stage we give to students only a small kernel of instructions (those related to the construction of cycles and alternatives) since we want to drive them rapidly towards an active use of computers putting off the refinements to the discussion of specific problems. In that way students can start programming very soon, trying by themselves to implement numerical methods in order to solve concrete mathematical problems.

Table 1. New topics added to the standard program

First year

INTRODUCTION TO THE LOGIC OF PROGRAMMING

(cicles, alternatives, flow-charts) .... 2 h

BASIC .... 4 h

COMPUTING ERRORS

(rounding, loss of significant digits) .... 1 h

GRAPHS OF FUNCTIONS AND GRAPHIC COMMANDS OF BASIC .. 2 h

METHODS TO LOCATE ZEROS OF FUNCTIONS

(bisection, secant, Newton's methods) .... 2 h

NUMERICAL METHODS FOR COMPUTING DEFINITE INTEGRALS

(rectangles, trapezoidal and Simpson's formulas) .... 2 h

Second year

DIRECT METHODS FOR SOLVING LINEAR SYSTEMS

(Gauss, pivoting) .... 3 h

SURFACE PLOTTING .... 4 h

METHODS FOR INTEGRATING ODE

(Euler, modified Euler, Taylor, Runge Kutta) .... 6 h

METHODS FOR NONLINEAR OPTIMIZATION

(gradient, projected gradient) .... 4 h

During this activity they encounter many problems of programming and little by little, with the help of a tutor, they discover all the triks becoming experienced programmers. Students work in groups of two or three writing their own programs and solving exercises: this organization helps the exchange of informations, accelerates the development of their programming skill. and contributes to make the entire group more homogeneous, as it has been shown by some tests (see the last section). This active programming work is also crucial to make students understand power and limits of numerical methods and to get them used to the application of computers as a research tool starting from their first year at university.

The choice of the language to be adopted is not very important in this direction and BASIC is probably the best choice if you want to minimize the number of prerequisites, considering that many students already know it.

Notice that during *working sessions on computers* the emphasis is mainly on mathematics rather than on programming. The goal is to have a program in order to test it on many different examples investigating the role of the set of hypotheses on the final result. To this end it is more important to try examples on which the program fails instead of easy examples on which it runs properly. Once again "computing errors" are used as a starting point to improve students understanding of mathematical properties. The presence of a tutor is essential to guide these investigation towards real mathematical problems suggesting tests and examples in order to show limits and benefits of the methods that have been implemented.

Table 2. A new teaching itinerary

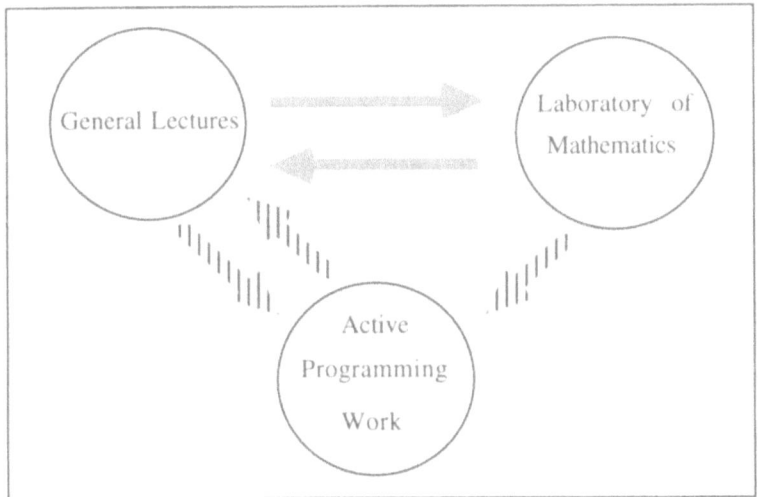

This interaction between the tutor and the groups is particularly important during the first year and much of the responsibility for the success of the programming activity is on his shoulders.

The *final examination* at the end of the first year consists in the solution of a a specific problem using the programs that they have developed during the working sessions. The problem is constructed in such a way that it does not fit exactly into the hypotheses of any method, then a clever adaptation of the problem and/or program is required in order to solve it. The students are requested to give the numerical results *and* to explain why they should be true.

At the end of the second year the final examination consists in a project . The subjects are quite advanced and a little beyond the contents of the course so that good knowledge of mathematics and programming skills are necessary to work out the project. Just to give an idea let us mention some titles contained in the list: graphical representation of curves in R^2 and R^3, numerical integration of systems of ordinary differential equations in R^2 and R^3, numerical solution of the Poisson equation on the square using finite differences, approximation of periodic functions by means of Fourier series and graphical representation of surfaces and manifolds.

The students where requested to develop a program, to write down a brief report explaining the methods they have used and to comment on the results obtained using their programs on various tests.

3. Further Developments

In the last two years some changes have been introduced in the experiment described above. The first one is motivated both by the necessity of a good framework for the programming activity and of a reasonable speed in computations. In the course of Advanced Calculus the study of complex models often requires quite long and complicated programs. Even if in principle these programs can be written in BASIC, this is not a good idea since in this language all variables are global variables and this make very hard to build routine and structured programs. So the best thing to do is to switch to a more advanced and structured language such as Pascal: in particular the Turbo-Pascal version is very appropriate for its simplicity and that is why it has been adopted. We must emphasize that *the choice of the programming language* is essentially related to the local situation in terms of equipment, to the objectives that you want to reach and to students background . Actually PASCAL and BASIC seem to be natural choices since they are the most popular general purpose languages and provide graphic facilities. In our experience, anyway, the switch from a language to another can be very rapid if students have a sufficient background. Now PASCAL has been adopted from the first year and only two hours have been added to the previous time-table (Table 1).

As we said before, even during the working sessions on computers the emphasis is on mathematical problems more than on programming. None the less the development of useful and effective codes is very time consuming even when the problems are simple. In order to minimize the hours of programming necessary to reach significant results we decided to write down a set of PASCAL procedures, a sort of *blackbox*, which could be included in students programs. This blackbox solve typical problems such as input verification, graphical representation of data, management of different areas on the screen, choice of items from a menu. The student need only to write down the mathematical kernel of his programs leaving to the blackbox procedures the management of the input/output . As a result programs developed by students are more homogeneous and clean so that thay can be used by others than the authors, moreover they can be linked together to form a library of interactive programs to be used for further investigations. This is the first step of the BIP (Biblioteca Interattiva di

Programmi) project whose final goal is to construct a library containing programs for a wide range of subjects related to the courses of Calculus and Advanced Calculus.

Another useful feature contained in the blackbox is the possibility of saving images on a disk so that any student can build an archive of examples which can be discussed with his collegues and/or teachers. In this way the discoveries made by one student can pass very quikly to the others accelerating the learning process.

The experience made in the courses of Calculus has now been extended to other courses such as Rational Mechanics, Probability and Statistics where computers are used essentially for demonstration. The development of the blackbox perhaps will modify this situation simplifying the organization of active programming sessions also for these courses.

Other (very few) experiments have been made in the courses of Algebra where the introduction of computers is a more difficult task since the object of this course is the study of abstratct structures (groups, rings, relations, etc.). In this course the aspect of simulation and visualition must be abandoned and the tentatives which have been done are concerned mainly with applications of packages for algebraic manipulations. In one case teachers have written programs to work in the ring of polynomials. The common opinion is that traditional languages for scientific computing are of very little help when applied to symbolic computing and further advances will be strictly connected with the diffusion of packages for algebraic manipulation now available also on personal computers.

Another course in which the application of computers as a teaching tool has been very limited is the course of Numerical Analysis. This is strange since computers have been traditionally used in this course since a long time ago and students are often requested to develop FORTRAN routines for the algorithms presented during lectures. Nonetheless, hardly any application of computers to visualize concepts and algorithms has been done. The development of the package MATRIX [9] was the only attempt and its goal was to find an effective way to illustrate some algorithms for linear algebra. The programs are conceived to represent the modifications on the matrix structure at different steps of an algorithm. Just to give an example let us describe the program which shows the solution of an algebraic linear system by the the Gauss elimination method.

The matrix elements can be represented either by circles with ray proportional to their absolute value (graphical representation) or by their numerical value. At each step a variable is eliminated in the system and at the end of the process the matrix is transformed in a triangular matrix having 1 on the principal diagonal.

Once the variable i has been "eliminated" the corresponding row and column remain the same till the end of the process. Figure 3.1 shows what happens on the screen when this algorithm works. The horizontal and vertical lines indicate the rows and the columns which have been already processed, i.e., which will not change in the following steps. Other programs contained in MATRIX use similar ideas to give a graphical representation of some iterative methods for algebraic linear systems (Jacobi, Gauss-Seidel, relaxation) and for the search for eigenvalues (Jacobi, Householder) (see Fig. 3.2).

58   M. Falcone

METODO DI GAUSS-RAPPRESENTAZIONE GRAFICA DI A

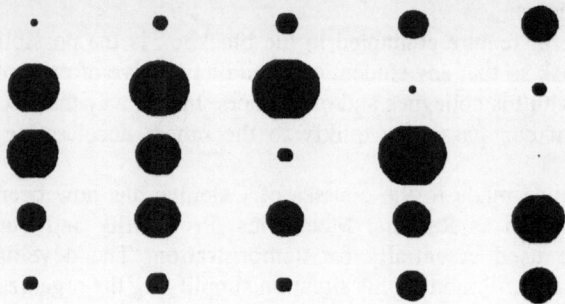

METODO DI GAUSS-RAPPRESENTAZIONE GRAFICA DI A
2 ^ITERAZIONE

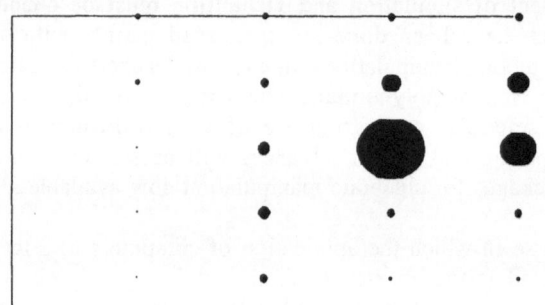

METODO DI GAUSS-RAPPRESENTAZIONE GRAFICA DI A
4 ^ITERAZIONE

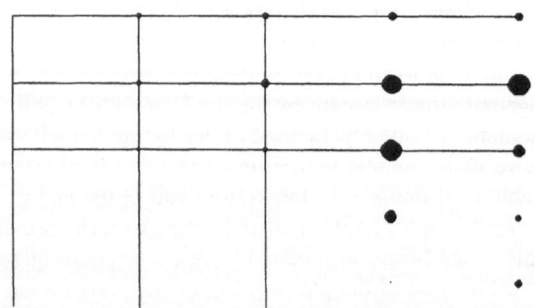

Fig. 3.1 Different stages of the Gauss elimination algorithm

RAPPRESENTAZIONE GRAFICA DI A
MATRICE SCELTA

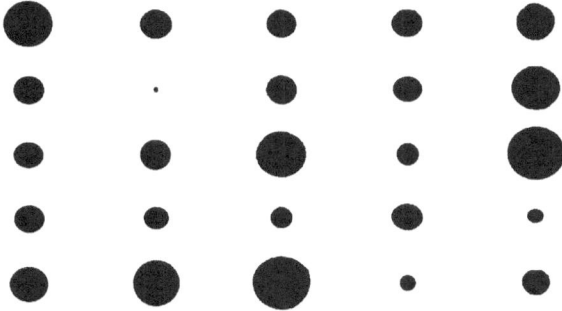

RAPPRESENTAZIONE GRAFICA DI A
15 ^ROTAZIONE

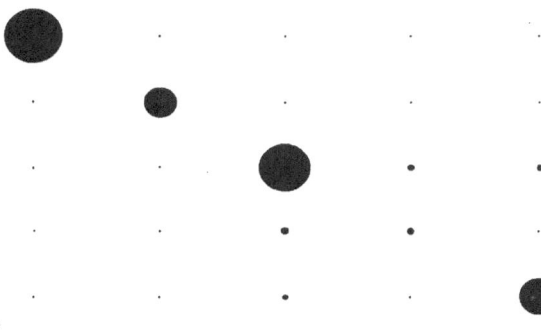

RAPPRESENTAZIONE GRAFICA DI A
30 ^ROTAZIONE

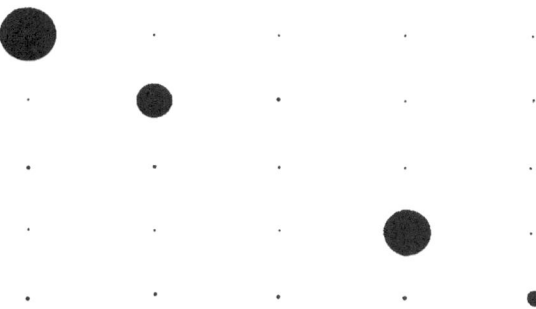

Fig. 3.2 The Jacobi algorithm searching eigenvalues

4. Conclusions

Even if the experiments are still going on it is time to fix, at least, some partial conclusions. An interesting element to *evaluate the impact of these experiments* on students learning has been a questionnaire submitted to all the students in Calculus.

```
- Number of students who answered the questionnaire:  152
    Students who own a computer:              36.8 %
    Students who intend to buy a computer:   47.3 %

  - Computer Group Students (CGS):   67   (44 %)
    Standard Course Students (SCS):   85   (56 %)

  - Marks :
              General average mark    4.13 / 9
              CGS          "       "       4.71
              SCS          "       "       3.67

  - Students who passed the test  (more than 4 right answers):
              Global result     40.8 %
              CGS               53.7 %
              SCS               30.6 %
```

Part of this test was made up by nine mathematical questions (see [3], [4] for the text of the questionnaire and for a more extensive discussion of the results) since the main objective was to know if the working session on computers can give to students a deeper insight in the whole subject. The test has been proposed both to students who have participated to the working sessions on computers (the CGS group) and to those who have not (the SCS group). The more relevant results are summarized in the table above. As it can be seen, the difference between the results of the CGS and those of the SCS is quite relevant and, what is more inportant, it has been almost the same on any question of the test (see [4]).

As a general remark on what we have experienced, *programs developed for demonstrations* in mathematics are more valuable for students if they are not so sophisticated that typical computer errors disappear. A student must be aware of the limits connected with the use of computers in scientific investigation and programs must not cancel mathematical difficulties: an overflow is an overflow and must not vanish by means of a programming trick! It is also very important for the student to have a sufficient background in order to distinguish his errors from computer errors. This is why it seems very necessary to introduce *new topics* related to the quantitative approach *in all the basic courses*. This approach greatly differs from that of the coursewares since programs are not intended has a substitution of leactures, on the contrary the use of computers described in this paper support and complement lectures.

The *visualization of concepts* in mathematics is a revolutionary event which will change completely our way of teaching increasing the role of geometry in the whole learning process. Not only it will possible to refresh traditional lectures giving in "real time" examples and motivations but it will be also possible to treat new subjects and give to the students some ideas of the new frontiers of current reasearch in mathematics. This seems to be a very important point to motivate students from the very beginning of their university career. If this approach will be developed also in secondary schools, probably the number of students choosing a degree in mathematics will increase, inverting the tendency that sees in most industrialized countries the best students moving to engineering and computer science departments.

As everyone knows the range of *applications* of the mathematical science has greatly increased: engineering and physics are the traditional fields of application but also biology, physiology and economics now use sophisticated techniques of modern mathematics. A presentations of simple models, their simulation on a computer and a discussion of the mathematical methods necessary to study them is probably one of the best way to motivate students who look at mathematics more as a difficult topic than as a useful science. This approach would be particularly valuable mainly for students attending degree courses other than Mathematics.

The working sessions at the computer laboratory and the use of computers as a dynamic blackboard are two different situations in which students work together and exchange informations. This *group activity* is really new for students in mathematics and has shown to be very important since accelerates students learning and contributes to the global success of the course rising the average knowledge of the whole class. Students learn also, and learn a lot, putting their hands in a problem and trying to solve it by themselves: the activity in the computer laboratory is the occasion to encourage and develop this attitude. The main difficulty is to focus their attention on mathematics rather than on programming and in this respect the access to already made PASCAL routines has been a great improvement.

Finally, the emphasis on applications and on the relevance of visualization does not mean at all that the abstract and rigorous approach to mathematical problems should be abandoned in favour of simulations and heuristic reasoning. The goal of this experience, in fact, goes beyond the concrete solution of mathematical problems even if numerical analysis plays an important role in pushing to a constructive approach and giving new instruments to improve students' learning of abstract concepts.

References

1. Bacciotti, A., Boieri, P., Moroni, P. (1986). An introduction to linear differential systems with the aid of personal computers, International Journal of Mathematical Education in Science and Technology, 17 (1), pp. 31-37.
2. Boieri, A., Chiantini, L., Geymonat, G., Moroni, P., Scarafiotti, A. (1984). Personal computers in teaching basic mathematical courses, in The impact of information technology on engineering education, Proceedings of SEFI Annual Conference, Erlangen.
3. Capuzzo Dolcetta, I., Emmer, M., Falcone, M., Finzi Vita, S. (1988). The Impact of New Technologies in Teaching Calculus: a Report of an Experience, Intern. J. on Math. Education in Science and Technology, 19 (5), pp. 637-657.
4. Capuzzo Dolcetta, I., Emmer, M., Falcone, M., Finzi Vita, S. (1988). The laboratory of mathematics: computers as an instrument for teaching calculus, in I. Capuzzo Dolcetta, M. Emmer, D. Salinger (eds.), ECM 87 – Educational Computing in Mathematics, North-Holland.
5. Capuzzo Dolcetta, I., Falcone, M. (1990). L'analisi al calcolatore, Zanichelli, Milano.
6. Capuzzo Dolcetta, I., Falcone, M. and Picardello, M. (1985). Utilizzazione dei microcomputers nell'insegnamento della matematica nel primo biennio universitario scientifico, CEE Report of Contract SSV-84-397-I, Rome.
7. Cottet-Emard, F., Garcia, F., Rivier M. (1985). L'enseignement des mathematiques par les moyens informatiques en premier cycle universitaire, CEE Report of Contract SSV-84-367-F, Paris.
8. Dechamps, M. (1988). A European Cooperation on the Use of Computers in Mathematics, in I. Capuzzo Dolcetta, M. Emmer, D. Salinger (eds.), ECM 87 – Educational Computing in Mathematics, North-Holland.
9. Falcone, M., Italiano, R. (1988). MATRIX: un software per l'insegnamento dell' algebra lineare numerica, Progetto Strategico C.N.R. "Tecnologie ed innovazioni didattiche".
10. Hardy, G.H. (1940). A Mathematician's Apology, Cambridge Univ. Press.
11. Lax, P., Burstein, S., Lax, A. (1976). Calculus with applications and computing, Springer-Verlag.
12. Salinger, D. (1988). The Teaching of Mathematics in a Computer Age. In: I. Capuzzo Dolcetta, M. Emmer, D. Salinger (eds.), ECM 87 – Educational Computing in Mathematics, North-Holland.
13. Salomon, L, (1984). Weierstrass and IBM. Bull. APMEP 343, 231-245
14. Truesdell, C. (1984). An Idiot's Fugitive Essays on Science, Springer-Verlag, pp. 594-631

Computer Aided Proofs in School Geometry

Régis Gras[1] and Italo Giorgiutti[2]

[1]Professeur de Mathématiques à l'I.R.E.S.T.E. (Université de Nantes)
et chercheur à l'Institut Mathématique de Rennes.
Campus de Beaulieu, Université de Rennes, F-35004 Rennes cedex, France
E-mail: gras@univ-rennes1.fr

[2]Professeur de Mathématiques à l'Institut Mathématique de Rennes.
IRMAR Campus de Beaulieu, Université de Rennes, F-35042 Rennes cedex, France
E-mail: giorgiut@univ-rennes1.fr

Abstract. Proof in geometry problems is an essential feature of secondary school mathematical teaching. This presents difficulties that many may pupils fail to overcome. Experiments with 'intelligent' logicials for pupil-aid in geometry, are described here. They are based on research into mathematical teaching and pupil learning patterns, and provide information on pupil spontaneous reasoning applied to geometry.

1. Introduction

Geometry problems which require proofs can create a variety of difficulties in the teaching and learning of the deductive reasoning.

Pupils can have difficulty:
- in finding the appropriate mathematical arguments in the corpus of institutionalized knowledge,
- in expressing ideas appropriately with the formal structure and the appropriate vocabulary,
- in expressing themselves rationally and unambiguously, and, above all,
- in understanding the sense of the mathematical proof itself.

Teachers can have difficulty:
- in situating and identifying the type of error committed by the pupil (logic, semantics, syntax, formal structure, etc.), recording repetition of errors, and their context,
- in formulating hypothesis on conceptions which underly pupil reasoning and procedures,
- in creating situations which will favour emergence of pupil reasoning, and which will discourage innapropriate patterns.

We have worked along two lines. In both, modelling pupil behaviour has been a permanent priority. The computer has proved to be both an efficient means of revealing and analysing pupil concepts, and of modelling them.

First research axis

This is concerned with error detection in simple one-step geometry proofs, and seeks to promote pupil understanding of the sense of the mathematical inference (12-13 year-olds). This work is briefly outlined in Sect. 1.

Second research axis

This is a study of pupil-exploration of geometrical figures in more complex situations (13–14 year-olds). We have attempted to promote pupil heuristic processing and pupil understanding of the sense of the proof and of the 'social and cognitive contract' or 'agreement' associated with this. This work is summarized in the following paragraphs.

2. The Basic Building Units of a Logicial Which Proceeds by Step-by-step Proofs

2.1 Preliminary Observations on Pupil Works (mainly with paper and pencil)

They showed certain consistencies in their behaviour, when presented with a simple inference which had to be completed, when one of its three terms was missing. These were:
  • difficulties in use of the formal vocabulary:
    since, then, because, therefore ... ,
  • symmetrical reading of the theorem,
  • choice of one of the three terms in the inference from:
    formal indicators (structure, rhythm ...)
    semiotic indicators (words, letters, symbols ...)
    semantic indicators (near sense ...).

2.2 Building a Logicial (Larher-Gras, 1989)

This structurally simple logicial can be used in secondary schools on nanonetworks with a microcomputer as server and is made up of three units.
  • *Pupil unit:* The pupil has numbered lists of facts and theorems. He has to solve a set of exercises, each of which involves a single inference, and which together may make up a complete proof. Each exercise presents the pupil with one or two omissions in the sequence:
    hypothesis(es)   theorem   conclusion
    The pupil has to replace the omissions with one or several facts ('*and*' and '*or*' can be used), or a theorem. Example:
    hypothesis1 and hypothesis? ... – theorem 5 – conclusion? ...
  • *Teachers unit:* The teacher composes the exercise-set, depending on the pupil patterns he wishes to encourage and follow for his own assessment. The

teacher identifies i) general errors (e.g., absence of a hypothesis, inversion of reasoning, from hypothesis to conclusion, treatment of the theorem as its reciprocal), and ii) so-called non-standard errors, i.e., those specific to the exercises (e.g., change in instantiation of the variables).
* *The evaluation unit:* Each pupil result is stocked, answers are totalized with counters (standard or specific). They are summarized in tables which pair pupils and response patterns and which can be analysed.

2.3 Example

In June 1989, we analysed a set of exercises, concerning the concept of central symmetry, for a French 5th form (British 2nd form, 12–13 year-olds). Certain consistencies in pupil patterns emerged from the classification of answer types and patterns:
* pupil success,
  instantiations were preserved , but expected relationships changed (*'parallel'* became *'equals'* or *'equals'* became *'parallel'*). The relationship between hypothesis and conclusion was based on symbolism,
* instantiations changed, but relationships were preserved. The relationship between hypothesis and conclusion was based on the sense.
* instantiations and relationships changed.

We stress that our hypotheses do not naively suppose that there is a simple transition from one step to *multi-step* reasoning, nor even that the latter necessarily proceeds from the former.

3. Hypotheses Concerning the Role and the Status of the Figure in Geometry Problems

Various observations in French 4th, 3rd, and even 2nd forms (British 3rd, 4th, and 5th forms) made in 1984 attempted to answer the following questions.
   1) How is a geometry problem initially perceived, at the beginning of a proof?
   2) How does the pupil get started? What are the means to which he can appeal?
   3) How is the figure exploited? Does it allow the pupil to anticipate the conclusion?

3.1

For the pupil, **a geometrical figure** is seen as the **single and objective expression** of the initial statement. Since the statement is almost invariably the teacher's (closed) creation , the pupil feels that it is inaccessible, and that it does not, in any way, belong to him. Consequently, he does not feel able to make any linear additions, other than those that may be indicated in his text-book. Since the pupil cannot intervene personally, he will neither assimilate the figure as part of his intellectual capital, nor relate it meaningfully to questions which refer to it.

3.2

In other words, the figure **is simply perceived and read subjectively,** or anyway in a manner which different from that of the teacher. Thus, the different implications of *"specific"* and *"general"* properties, of a *situation* or a *theorem*, are not fully appreciated. In particular, what is merely contingent and what is actually pertinent and invariant remain confused. Consequently, it is not the way in which the pupil percieves the figure that will enable him to solve the problem. We have seen this in an experiment in geometric construction, in which the demonstration of a relationship between segments went unnoticed by pupils, though evident to us. An element of a figure is not a general trait (i.e. a pertinent and discriminatory sign). Teachers should remember, for example, that a pupil does not see a square as a special case of a parallelogram, but as an autonomous geometric entity in its own right.

It is also true that, for the pupil, the logically acceptable **relationship between the signifié and the signifiant** – i.e., one of the codes which representes the signifié, can remain **ambiguous** for a long time. The univocity of the objects of his attention (the concepts and symbols, or the words which designate them), and the homomporphism of the relationships between them are not seen to be stable.

This ambiguity is a didactic obstacle. Sometimes, for example, the 'reading' of a figure or a graph itself is taken as a proof (in particular in a counter-example, sometimes it remains suspect, and has to be verified in other ways (Gras, 1988).

3.3

The degree of certainty of a conjecture stimulates elaboration of the proof. If this certainty is too great, then it removes any need to prove anything (because a proof is seen as trivial or superfluous). If it is too low, it carries with it a risk of error which is seen as excessive. The formal **proof would then seem to be more a contractual agreement between a teacher and his pupils.** The kinds of proof that the pupil willingly and spontaneously uses vary in their expression, and not all of these are necessarily 'intellectual'. Moreover, the pupil (like the adult), who is not himself persuaded by his own perception and intuition, will resist any conviction that one tries to impose through formal proof, however rigorous this may be (Balacheff, 1989).

3.4

The frequent confusion between hypothesis and conclusion would also seem to be more a linguistic than a logical problem (understanding terms such as *"if......... then"*, *"because"*, *"since")* . Moreover, a proliferation of pupil hypotheses, and inadequation between these and the sought-after proof, show the pupil has not understood that the status is related to state of development of the problem. A study of three French 4th forms (British 3rd form) in 1984, showed that all the pupils who had perfectly identified the hypotheses pertinent to a sought-after conclusion, had either correctly established the proof, or had shown a rational, and therefore unambiguous, understanding of the problem (true to 98.5% implication with the R. Gras Index).

3.5

In problems whose solution requires a proof, **the didactic contract,** established implicitly, permanently and reciprocally between the pupil and the teacher, is the foundation of the didactic relationship. This can be seen when that relationship is perturbed. For example, when a pupil does not understand the sense of what he is being asked to do, nor what kind of answer he is expected to give. This is particularly so for French 4th form pupils (British 3rd form). who still, though increasinly less nowadays with new teaching procedures, discover new properties in geometrical figures, which had previously been perceived purely 'functionally'. They progress from *'descriptive'* geometry, familiar in the context of drawing and measurement, to a 'deductive' geometry. Their reasoning, referring to rules which are not explicit, may end up with quite incoherent imitations of the teacher's proofs. The pupil has understood neither the role nor the syntax of the proof, but has accepted the proof as such, simply to satisfy the teacher.

From the teacher's point of view, **teaching a proof is different from teaching** what may be described as **'cultural objects',** or institutionalised hability. Seen from the pupil's point of view, the teacher:
- knowing **otherwise**, can decide on the appropriate objects of knowledge, and the right roads to understanding (*"Watch what I do, then do the same"*)
- **judges the validity of pupil-reasoning,** modifying, accepting, or rejecting his arguments, referring to implicit requirements which he alone decides on. The pupil may well present a rigorous proof, and following an example, but this will be criticized, or even rejected by the teacher, because it lacks clarity or elegance, as judged by infdefiniable requirements.
- **proceeds by sufficient conditions** within explaining the solution search (backwards) and requires of pupils a deductive presentation(forwards).

In summary, pupil response to geometrical problems with proof involve varied and complex reasoning patterns, of which all are by no means specific to the geometry itself. These can be related to:
- **social factors** – ability to convince, to test, to reason,
- **psychological factors** – ability to convince oneself, resolve conflicting options,
- **cognitive factors** – ability to acquire knowledge about geometic objects and exposure methods which is more or less evident .

4. Our Hypothesis Concerning the Role the Computer and Our Aims in Using It

In order to established from our observations, and some teaching practises, the elements of an educational software program, we have developed certain hypotheses, and attempted to control some didactic variables. We have chosen to work on real pupil activity (affine geometry), traditionally set for French 4th or 3rd form pupils (British 3rd and 4th forms). This work has followed two lines.

- **The first line is heuristic.** This seeks to allow the pupil to assume the problem situation in all its specificity, to speculate, and to develop his solution through a set of subproblems, or at least in rationally interdependent steps. This method cannnot be assimilated to techniques generally described under *"problem-solving"*, or to algorithmic learning, to which pupils in difficulty may all too readily turn.
- **The second line is socio-didactic.** This seeks to give meaning to the proof by establishing an agreement between the pupil and the computer about statements concerning the goals which are to be achieved. It also seeks to develop the logico-linguistic means to this end. We encourage the confrontational context getting pupils to work in pairs on the same same problem. We hope that in so doing the pupils proof task overstep the simple semantics procedure.

4.1 Construction of the Figure and Its Exploration

Wherever necessary, we oblige pupils to set up their own conjectures about the situation with which they are confronted, by getting them to work actively on the figures. This is because preparation of the figure is often perceived by pupils as a task in its own right. The figure itself is then seen as something foreign to, and independent of the search for, and development of, rational articulated reasoning, necessary to appropriate proofs and hypotheses. We hypothesize that we should favour **backwards reasoning**, and that it will better to work back from a specific goal, to previously acquired knowledge.

A computer system (grapical table + IBM PC) designed by I. Nicolas, IRISA, (Nicolas, 1989; Communication by L. Trilling) allows pupils to:

– compare the figure they have constructed and its underlying hypotheses, with logical implications of the latter as translated by the computer system,
– to change their hypotheses, and to animate the figure locally.

In our logicial, the "DEFI" logicial, the didactic variables, in other words those that the teacher controls, and which can affect pupil reasoning, are chosen from options, which the teacher can select from the computerized system(these options are described in Sect. 5.2).

A series of questions, designed to stimulate pupil reflexion, allow him to explore the figure, and prevent him from '*stalling*'. They also allow him to acquire, by his own efforts, a certain understanding of the relationships between the different elements of the geometric situation with which he is confronted. The next phase allows the pupil to distinguish clearly between the '*abstract*' geometric object, its graphical representation, and between conjecture and mathematical proof respectively.

4.2 The Proof

Although the logicial offers pupils tutorial guidance, or a '*helping hand*', it does not provide the answer to the problem, but rather:

- tends **to delegate the problem** to the pupil, encouraging him to establish an explicit agreement with the machine. It is this agreement which should make his development of the proof meaningful, and which he should honour by respecting the accepted logical rules. The agreement should de-personalize the didactical relationship between the pupil and the problem.
- increase the **stake of the proof** because the logicial does not indulge in, and will not accept, any rule-breaking or sophistry; also because the answer to the problem is neither evident, nor trivial, nor unbelievable (an important didactic variable).
- confronts **backwards and forwards reasoning**, so avoiding an impasse at the beginning of a deductive sequence, and losing sight of the problem's significance.
- distinguishes between the *signifié* and the *signifiant*, through the dialectical opposition of two logicial viewpoints, in which declarations about perceived relations and properties, conjectures, statements and mathematical proofs will be confronted.

The 'operationalization' of these objectives will be considered later. We believe that the computer is a particularly useful tool, which can be employed in the classroom to follow individual patterns of reasoning, and individual rythms. It also seems to us to be an incomparable means of describing and analysing reasoning patterns, pupil resistances and difficulties. It can be the teacher's ally, both in teaching and learning, and in teacher assessment of pupil procedures.

5. The Creation, Experimental Use, and Assessment of the DEFI[1] Logicial

5.1 The Logicial's Structure and Use

The logicial has been progressively modified over successive experiments, since the 1985 Prolog prototype, including new functions to improve its didactic value, and greater delegation of the problem to the pupil.

Currently, DEFI, written in Pascal, is presented to the pupil in the following modules.

- a 'DEFI' module, describes the use, and management of the logicial, with instructions for navigating within and between the different units and menus.
- Two main modules, made up of dialogue sequences.

– the 'figure exploration' module, with a mini-expert-system (the figure traced on paper). This involves a relatively sequential set of questions, which backwards approach hypotheses taking conclusion as starting point. Pupil responses can be divided into classes:
  – conjecture resulting from exploration of the figure,
  – declaration of ability to provide the proof, recorded in a summary,
– a proof module.

The pupil states the condition that he wishes to prove, and specifies the appropriate theorems, together with their relevance to the facts (or hypotheses) of the problem in its currently state.

To this end, he has access to a computer file containing the theorems pertinent to the field of interest. He also has a file containing specific templates of the type *"the point ...... is the middle of the segment ......"*, or *"the straight lines ..... and .... are parallel"*. The logicial leaves the pupil free to organise his own strategies, but will return him in case of failure, to a new proof, or a new exploration of the figure.

The complete proofs for a given problem are in fact an ordered, logical sequence of one-step proofs.

• a dialogue can be interrupted at any moment, to access various pull-down menus. These are:
  – a *'navigation'* menu, which allows an activity to be changed (e.g., choose another problem, begin the problem again in case of error, etc.)
  – an *'information'* menu, which presents the statement of the current problem, the overall summary in which the diferent conjectures are recorded and updated, the elements which have to be proved, the properties which have been proven, and the state of advancement of the proof.
  – an *'aid'* menu.

5.2 The Didactic Variables

Variables are didactic (Brousseau, 1986) when they are responsible for qualitative changes in pupil procedures and strategies. They are pertinent in that they can account for for results, and their control can affect pupil performance. Amongst the variables that we examine below, certain are clearly didactic. Nevertheless, for these, and for others, it still remains to see whether they behave so:

• concepts and relations between them, in a given problem (there are currently 9 problems available, an example is given in Appendix 1),
• order of construction of the relevant figure (they do so in the present version of the logicial),
• the degree of confidence in the final and secondary goals,
• the number of steps which are involved in the proof and the rules of inference which are accepted (no proof by reasoning from the absurd),
• readiness to conjecture, to declare one's ability to provide a proof, to consult, to specify the aims of the proof,
• access or not to *"navigation"* menu ,

- number, type and formulation of the theorems in the tool file (10 available in the present logicial)
- access to the tool file.

5.3 Experiments

5.3.1 Methods

Throughout development of the logicial, the different versions were tested, more or less empirically, the method currently being:
- **2–5 Macintosh** computers, with pupils working in pairs, freely established, without any individual task-attribution,
- **1–3 observers** (teachers-researchers) per workstation
- **observations** which attempt:
  - to specify points of particular importance for observation
  - to describe pupil activity in three categories: the stages through which each pair of pupils passes, the relationships between members of the pairs (the vocabulary used in verbal exchanges and its influence on their actions) and the written traces, either on paper or on the screen.
- **individual writing** up of the problem afterwards, and at home,
- **a short report** from each observer,
- **a general assessment** of the experiment.

5.3.2 Different Versions of the Logicial Have Been Tested

These experiments took place in the UER de Mathematics and Informatique (the University Department of Mathematics and Computer Sciences) with a French 4th form (British 3rd form), at the Lycée de Villejean (Rennes)with a French 2nd form (British 5th form), and at the Collège des Gayeulles (a secondary school, which takes children between French 6th and 3rd forms – after which pupils go to a Lycée), with a French 4th form. **The general results of observations on the last group are presented here.**

The experiment was carried out at the end of the school year, with eight pupils from a 4th form. Generally the pupils in the pairs, who were volunteers, were not, in any way, *'brilliant'* in maths. Some of them might even have been considered as being *'in a delicate situation'* as far as geometrical reasoning went.

For the logicial to be adopted and used efficiently by the pupils, they must understand and accept a new didactic agreement. This may, in some respects, conflict with that which, they have accepted previously in class. For example, the preliminary exploration of the figure may, at first, be felt to be slow, trivial, and invalid, because it compounds the conjectural aspect (the fruit of observation), and the deductive aspect *("can one affirm something without a proof?")*. The ambiguity is dispelled once they begin to use the *"demonstration"* module.

In the same way, **the constraints of a one-step proof are not readily understood and accepted**. The intuition of an inductive approach usually contradicts it. Nevertheless, the observers found that, during the proof, the constraints were beneficial:

– they favoured rigorous construction of the proof, and, in particular, analysis of the theorem's text, in terms of hypothesis and conclusions, and also of its pertinence, relative to the hypotheses emmitted and conclusions drawn.
– experience was reinvested in subsequent problems over the short term.
– above all, they decreased the load the implicit conditions impose, and so blur the agreement of exposition of the proof.

The 'Summary' File seems of great interest:

– it favours establishment of an explicit agreement, in harmony with the socio-cognitive aims of the proofs. The agreement is not tacit, but rational and explicit between two participants (one, the pupil, is real, the other, the micro-computer, is fictive). The pupil is thus obliged to judge his actions himself.
– the logicial is also constantly updating the development of a solution of the problem. The differences between the status of conjectured and proven properties is clear.

More generally the communication established with the logicial, without any loss of rigour either in content or form, and the relationships between members of working pairs clarifies the nature of the agreement governing, although this may be implicitly, geometric proof.

Return to the 'Theorems' file clearly identifies its role of external tool, inside a theory. In fact, this file foreshadows and simulates the collective memory of institutionalized facts, on which one is allowed to draw (cf. the metamorphosis we have illustrated in the film *Reflets et Taches* produced by IREM Rennes).

Understanding of the disymmetry between hypothesis and conclusion is strengthened:

– when the pair receives messages such as *"such and such a property is not a hypothesis, or has not been proven"*,
– when the pair has tried to prove two steps and where the hypothesis for the second, is in fact conclusion to the first.

We must stress here the importance, recognized by the pupils, of **the immediate sanction for any mistake** made during a proof. This is far more rapid the lag in normal teaching (corrections on the board, or on a test paper). This was expressed in the reports of the eight pupils. The contents of these reports was, for the teachers, surprisingly better than those prepared under standard teaching conditions. This confirms the positive influence of this on-line guidance during exploration of the figures. It also confirms the interest of the distinction made between the properties which are given, and the properties which are deduced.

Finally, **the working in pairs** seems to have favoured:

– better understanding of the communication agreement made between the pupils and the micro-computer ,
– development of critical aptitudes of pair members, who, through their inter-action, may anticipate the objection that the logicial would have raised.

Nevertheless, we must also note that:

– despite the fact that the activity is positive, it is costly in time,
– it is impossible to know whether the effects of the experience reinvested in the long term, in traditional work with pencil and paper.

6. Developments and Perspectives

The assessment of the logicial in its present form underlines, amongst other things, two difficulties to which we wish to draw attention.

1. First, we note, once again, that the pupils come to grief on the obstacle of precise identification (pertinence, minimality, ...) in the hypothesis(es) associated with an affirmation, and which still remains to be proven. One might hope for some improvement in learning, from the 5th form (British 2nd form) and perhaps from the 6th form level (British first form),one would also hope that such improvement might allow pupils to make better choices, and to order these pertinently, when faced with an explicit list of affirmations and theorems with three elements:

Hypothesis
Theorem of the type – if p then q
Conclusion

1. We shall therefore continue our research along the first line (cf. Sect. 2), extending the logicial's possibilities to assembling of several steps in a proof for a same problem, and also so as to allow the pupil free access to any level of that proof.

2. We remain interested in the second approach. The tutorial unit imposes certain forms of written proof, but in which virtually no original textual production is required on the part of the pupil. The usefulness of this kind of help may then remain limited, if the pupil, subsequently left on his own, is unable to formulate the terms of his proof. However, any spontaneous written production reflects a individual presentation and language, that the teacher must take account of, and that the didactic must be able to register.

In future we shall seek to increase pupil's written participation by developing the analysis of its expression, independently of the logical coherence of the proof per se.

We also envisage extending the scope of the problems which can be treated to problems of metric type, and to development of a PC compatible version of the logicial.

Our future research program will therefore include, didactic reflexion, developments of tools, experimental assessment of our hypotheses, including those concerning didactic variables.

The traces that the pupils have left during their exploitation of the DEFI logicial, i.e., the order in which they have chosen options, messages, and help texts, will be a rich source for didactical analysis. The results obtained so far encourage us to follow up research in towards rational incorporation of computerized aids in mathematics teaching. These results convince us that a teaching logicial is a team project involving not only programmers, but didacticians and teachers as well.

References

Ag Almouloud S. (1992): L'ordinateur, outil d'aide à la démonstration et au traitement de données didactiques, thèse de l'Université de Rennes 1.

Ag Almouloud S. (1992): Aide logicielle à la résolution de problèmes avec preuve : des séquences didactiques pour l'enseignement de la démonstration, RDM, Vol. 12, n°2.

Ag Almouloud S., Giorgiutti I. (1993): La modélisation de l'élève : le cas de D.E.F.I., Actes des 3èmes Journées EIAO de Cachan, Eyrolles, Paris.

Ag Almouloud S., Giorgiutti I. (1994): EIAO et didactique : le cas de D.E.F.I., outil didactique et d'aide à la recherche, RDM, Vol. 14, n°1.

Balacheff N. (1982): Preuve et démonstration en mathématiques au Collège, Recherches en Didactique des Mathématiques, 3.3, 261–304.

Balacheff N. (1985): Etude des processus de preuve en mathématiques chez des élèves du premier cycle de l'enseignement secondaire, thèse de Doctorat d'Etat, Université de Grenoble.

Brousseau G. (1986): Théorisation des phénomènes d'enseignement des mathématiques, thèse de Doctorat d'Etat, Université de Bordeaux I.

Chevallard Y. (1985): Pour introduire à l'ingéniérie didactique à composante informatique, Rapport sur l'Université n° 20 de Luminy.

Gras R., Boisnard D., Allen R., Nicolas P., Trilling L. (1987): Gestion informatisée de problèmes et de démarches liées à leur résolution, La Nouvelle Encyclopédie Fondation Diderot.

Gras R. (1988): Une situation de construction géométrique avec assistance logicielle. Recherches en Didactique des Mathématiques, vol. 83, 195–230.

Greco Didactique et Acquisition des Connaissances Scientifiques, Informatique et ingéniérie didactique, Rapport d'activités 1984-85 du sous-thème 2 du thème 3, des IREM. de Paris Sud et Rennes, du CATEN., de l'INSA., de l'IRISA. et de l'IRMAR., IREM. de Rennes (mai 1985).

Larher A., Gras R. (1989): Le micro-ordinateur, outil de révélation et d'analyse de procédures dans de courtes démonstrations de géométrie. 13ème Conférence Internationale Psychology of Mathematics Education, Paris, juillet 1989.

Nicolas P. (1989): Construction et vérification de figures géométriques dans le système MENTONIEZH, thèse de l'Université de Rennes I.

Py D. (1990): Reconnaisance de plan pour l'aide à la démonstration dans un tuteur intelligent de la géometrie, thèse de l'Université de Rennes 1.

Appendix – Screencopies from a Run of DEFI

```
Information  Aide
 Enoncé du problème
 Bilan général        ▪

Problème n°9


On considère un triangle ABC et on désigne par D et E

les milieux de [AC] et de [AB]. Soit M un point quelconque

du plan et soient G et H les symétriques de M par rapport à

E et à D.

Démontrer que le quadrilatère BCHG est un parallélogramme.
```

```
Navigation   Infor
 Quitter
 Menu principal

 Autre problème


    Numéro du problème choisi: [9]

  ( Confirmer )
```

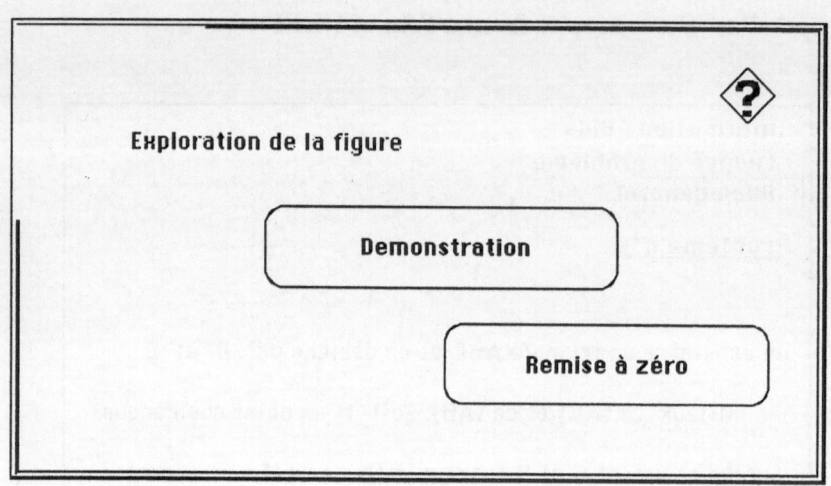

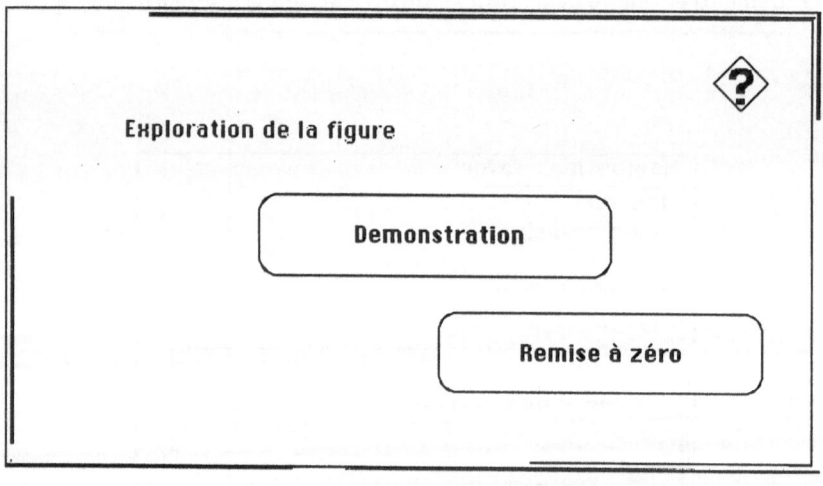

Sais-tu démontrer que

Le quadrilatère BCHG est un parallélogramme?

Oui   Non

Sais-tu démontrer que

BC est égal à HG ?

Oui   Non

Existe-t-il un segment dont la longueur

○ soit égale

○ soit egale au double

○ soit égale à la moitié

des segments [BC] et [HG] ?

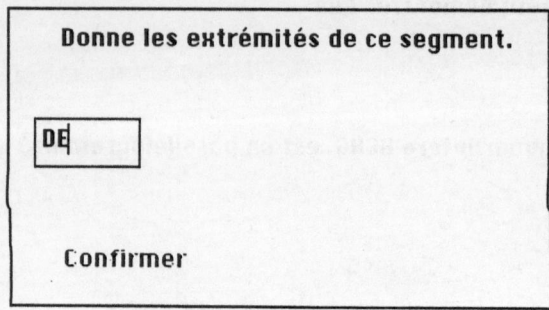

Donne les extrémités de ce segment.

DE

Confirmer

Information Aide
⌐ **Enoncé du problème** ⌐
Bilan général

tu dis savoir démontrer que:

La longueur de GH est égale à deux fois celle de [ED].

La longueur de BC est égale à deux fois celle de [ED].

Les droites BC et HG sont parallèles.

Tu as pu observer que : BC et GH ont même longueur

Exploration de la figure

Demonstration

Quelle propriété vas-tu démontrer?Choisis dans la liste suivante:

○ **les points ... sont alignés**

○ **Le point . est le milieu de [..].**

○ **Les droites (..) et (..) sont parallèles**

○ **Le quadrilatère .... est un parallelogramme.**

○ **La longueur de[..] est égale à celle de [..] .**

● **La longueur de [..] est égale au double de celle de [..] .**

La longueur du segment | BC | est egale au double de celle du segment

| DE | .

Confirmer

Quel théoreme vas tu utiliser pour cela? Fais ton choix dans la liste suivante:

○ Si les diagonales d'un quadrilatère se coupent en leur milieu, alors ce quadrilatère est un parallélogramme

○ Les diagonales d'un parallélogramme se coupent en leur milieu.

○ Un parallelogramme a des côtés opposées égaux et parallèles.

○ Si un quadrilatère a ses côtés opposés parallèles, c'est un parallélogramme.

○ Deux droites parallèles à une même troisième sont parallèles entre elles

○ Par un point il passe une parallèle et une seule à une droite donnée.

○ Un quadrilatère qui a 2 côtés égaux et parallèles est un parallelogramme

● Le segment qui passe par les milieux de 2 côtés d'un triangle est parallèle au 3˚ côté et sa longueur est égale à la moitié de celle de ce côté.

⇨ Suite

Sur quelles hypothèses t'appuies-tu? Choisis dans la liste
suivante:

○ les points ... sont alignés

⊙ Le point . est le milieu de [..].

○ Les droites (..) et (..) sont parallèles

○ Le quadrilatère .... est un parallelogramme.

○ La longueur de[..] est égale à celle de [..] .

○ La longueur de [..] est égale au double de celle de [..] .

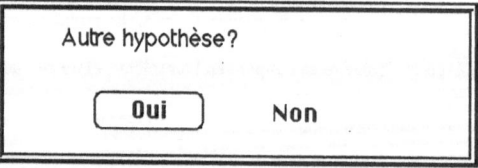

Le point │ D │ est le milieu du segment │ AC │ .

Confirmer

Autre hypothèse?

[Oui]   Non

Compare le nombre d'hypothèses que tu as donné avec celui des
hypothèses du théorème choisi.

Ok

Navigation   Information   Aide

Enoncé du problème   F

Quel **Bilan général**   i:

suiva   état de la démonstration

tu souhaites démontrer que:

La longueur de [BC] est égale à deux fois celle de [ED].

En appliquant le theorème suivant:

Le segment qui passe par les milieux de 2 côtés d'un triangle
est parallèle au 3° côté et sa longueur est egale à la moitié
de la longueur de ce côté.

En utilisant les hypothèses suivantes:

Le point D est le milieu de AC

A Cognitive Analysis of Geometry Proof Focused on Intelligent Tutoring Systems

Dominique Guin

ERES Département de Mathématiques, UM2
Place E. Bataillon, 34095 Montpellier cedex 5, France
E-mail: guin@math.univ-montp2.fr

The elaboration of an Intelligent Tutoring System in geometry proof requires a model of geometry proof problem solving. Actually, the development of these models is still directed by the system's processes and not by a study of human behaviour. Here we present elements of a cognitive and didactical analysis of proof in geometry that must be taken in account in such a model. This study was carried out with mathematic teachers (GIA: Artificial Intelligence group, IREM Strasbourg).

1. Two Distinct Activities in Geometry Proof

1.1 The Deductive Organization

To tackle proof in geometry, students have first to understand the *rules of the game*. Only later they will be able to discover a solution. So they have first to master the *deductive organization* [Duval 91]. In this activity, all the rules (definitions or theorems) and statements used in the proof are *given* to the students, they have to:

- put these statements *into order*,
- choose one rule,
- *check* on each step the *substitution* linked with the applied rule.

The first task requires understanding the different *operating roles* of each statement (the givens, the conclusion, intermediate conclusions, etc.).

The second task requires a *procedural* (and not only *declarative*) knowledge of the rules (definition or theorems), that is to say an *operating* knowledge. An efficient help in the deductive organization is given by the use of a *deductive net*, whose nodes are statements (their different roles are pointed out by different colours) and whose arrows are rules: this deductive net is a *checking tool* for the student.

1.2 The Discovery of a Solution

We will try to explain our "expert" behaviour: it seems that we use no explicit rules to guide our search. These rules do not allow us to find immediately a solution, but instead allow us to define a *plan of action* and then perhaps get a proof after making some fruitless attempts. To define a plan is to *imagine a way* between givens towards the target without giving all the steps. A plan is *not always successful*, it is not always a sequence of rules (definition or theorems) to be applied, but more frequently a sequence of subgoals to be reached. Imposing a *step by step* method may often be an *obstacle* to the discovery of a solution, this is a *too localized* vision of a proof. *Heuristics* are a help to find a plan. We must make our heuristics explicit and we must be able to *justify our choices* of a plan.

We make the assumption that to build a planning net could be useful for students to elaborate their plan. But it would be different of the deductive net: the nodes of this planning net would be subgoals and arrows would not correspond only to rules (e.g, they could be identifications of prototypical configurations for example and more generally heuristics).

It is essential to let the student test his plan *even if it fails*. In geometry proof, an expert does not systematically choose first a successfull plan. The student will explore his plan to understand *why* it fails; and we hope his second choice will be more *appropriate*. An *immediate feedback* on errors like in [Anderson & al. 87] would hinder any progress in this activity.

One notices different features in *planning nets*, one finds:
– arrows in two directions,
– different roles: statements, prototypical configurations, etc.,
– different configurations to be "recognized",
– heuristics which may fail,
– assertions without verification.

1.3 An Example

ABCD is a parallelogram. I is the intersection of its diagonals. E is the midpoint of segment [CB], F is the midpoint of segment [CD]. Straight lines (AC) and (EF) are secant on M. Prove that M is the midpoint of [EF].

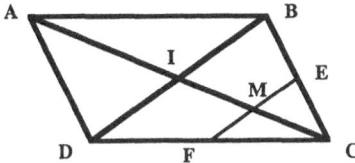

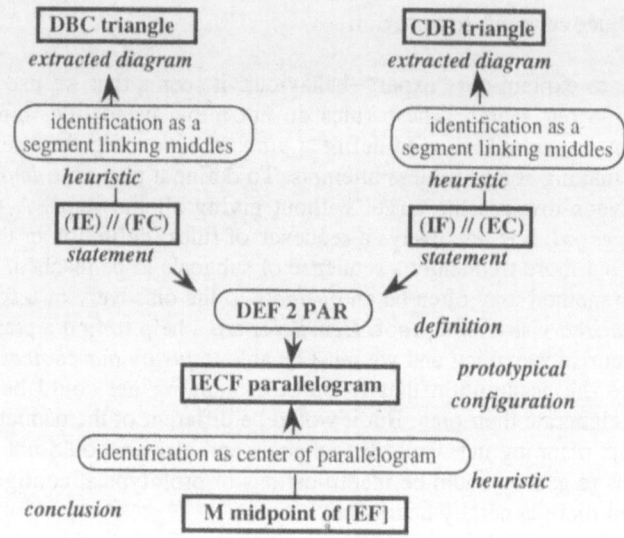

After the planning net, here is the deductive net:

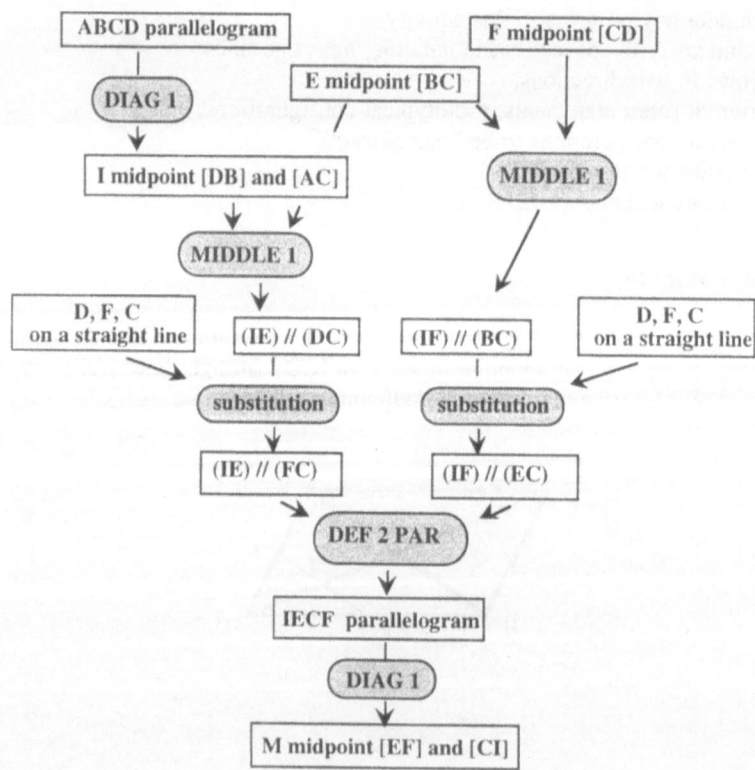

2. Which Impacts on Intelligent Tutoring System?

We will express some hopes for the *learning* module:
- the aim of an intelligent tutoring system for geometry proof is not automatic proof. It is focused on the *guided discovery* of a proof. It must accept a *partial solution* and provide help to complete it. It must be able to solve the problem in the *way students* do.
- such a system is able to *comment, explain* and *justify* its method [Nicaud & al. 88]: it is not sufficient to give the applied rule, the system has to justify its *choice* of rule.
- the student *keeps the initiative* within the domain possibilities: he can carry on with his plan even if it is not the best, even if it will fail.
- the ITS has to *fit* with the student *level*: we stress the importance of giving students an active role in making the representations of proof steps advance in accordance with students' understanding [Kaltenbach & al. 89].
- we suggest a structure enabling students to *separate difficulties* pointed out in the analysis of learning geometry proof working in a specific submodule.

This learning module would be composed of different submodules :
- a *knowledge basis* submodule working on different representations of definitions and theorems,
- a *representation* submodule emphasizing with a net the role of statements (translating problem understanding),
- a *diagram construction* submodule as Cabri-géomètre [Baulac & al. 88],
- a *diagram investigation* submodule linked with the terms of the problem to underline, on the diagram, roles of statements, prototypical configurations etc.
- a *planning* submodule (to elaborate a plan),
- a *deductive organization* submodule (to check a proof),
- a *proof* submodule (allowing work in subproblems in the order wished by the student).

We will detail the objective of some submodules (for more details see [Guin & GIA 89]):

2.1 The Knowledge Basis Submodule

Students have the possibility *for each geometric object* to work on:
- *different representations* of the object, with moving diagrams, for example, 4 *prototypical* configurations for the parallelogram:

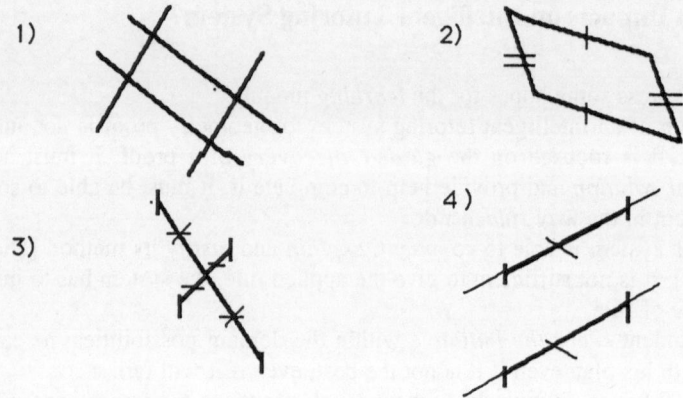

(These prototypical configurations come from a procedural representation of definition and theorems).

– *procedural* representations of definitions and theorems with a systematic study of converse statement even if it is an equivalence (the two corresponding rules are separately used). Such a representation underlines the different hypotheses necessary to apply a theorem, for example:

DIAG 2: Every quadrilateral with diagonals secants intersecting in their midpoint is a parallelogram.

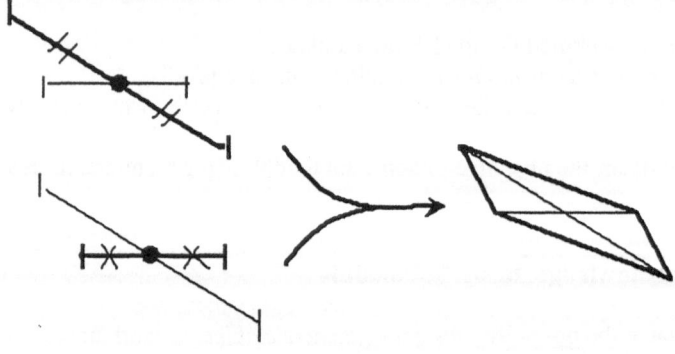

MIDDLE 1: The straight line going through the midpoints of two sides of a triangle is parallel to the third side. The length of the segment linking the midpoints of two sides is equal to the half of the length of the third side.

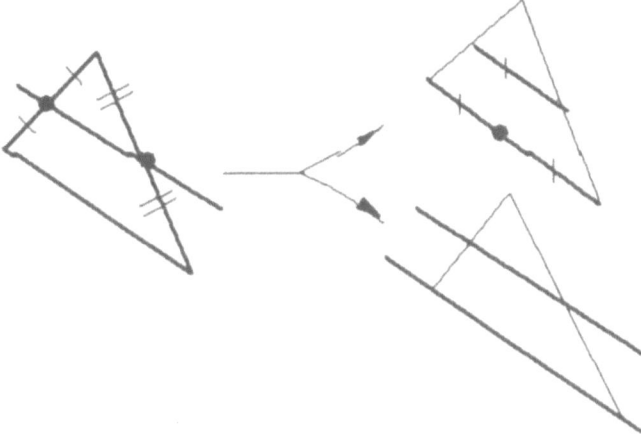

– *semantic net* of geometric objects, with specializations and generalizations, also underlines the procedural aspect of definitions, for example:

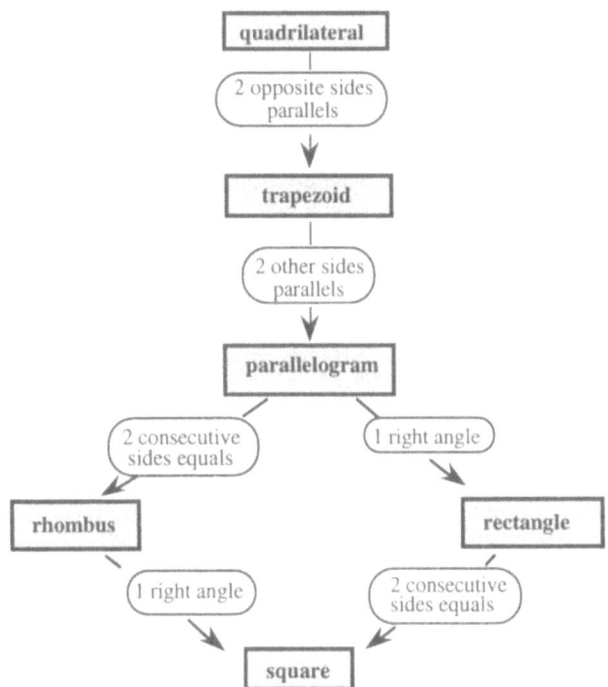

2.2 The Representation Submodule

This submodule should allow us to work with different problems to point out the role of statements. The knowledge organization should be presented by the student. The problem would not be given in a predigested form as in Geometry Proof Tutor [Anderson & al. 87] where roles are already provided. This submodule should provide a translation of problem understanding. We notice that, at this moment, there is no representation well adapted to such a submodule. Therefore existing software does not include this submodule which would be, in a way, an interface with natural language.

2.3 The Diagram Investigation Submodule

This submodule already exists in [Gras 88]. It should be developed to make explicit the relationships between the diagram and the terms of the problem to underline, roles of statements, prototypical configurations, etc., with the aim of bringing out conjectures for an outline of the plan. This submodule would be linked to the representation and knowledge basis submodules.

2.4 The Planning Submodule

The student should be able to elaborate a plan for a problem or a subproblem. ITS should have to fit with the student level:
- complete plan: ITS should not impose search to a student who has a solution, it should not ask for too many justifications when there is no difficulty,
- incomplete plan: ITS should accept an incomplete plan, and should provide helps to complete it,
- no plan: ITS should provide help to find a plan using heuristics such as:
  If you want to prove X is midpoint of segment [YZ], test:
  - an identification of a prototypical configuration DIAG 1 whose one diagonal is [YZ] and center is X,
  - an identification of a prototypical configuration MIDDLE 2 in which there exists a straight line parallel to another side that [YZ] going through X and the middle of the third side,
  - if X, Y, Z are on a straight line and X is on the perpendicular bisector of Y and Z.

The identification of an extracted or prototypical diagram is composed of two steps: showing the diagram, then checking the properties of an extracted or prototypical configuration.

Specific helps of this submodule would be:
- backward help (pointing out goal),
- forward help (pointing out given),
- knowledge basis with heuristics.

This submodule should be linked to the knowledge basis, to representation and diagram investigation submodules by helps, and the control of the consistency of the plan in view of the givens.

2.5 The Deductive Organization Submodule

Firstly, this module should allow us to work with different problems to point out the role of statements and to recognize if rules provided can be applied.

Secondly, it should allow us to check the deductive organization of a geometry proof whose rules are provided.

Lastly, it should allow us to check the proof whose plan was previously built. Specific aids of this submodule should be:
- rule help (procedural representation),
- compatibility between the chosen rule and the deductive step.

This submodule should be linked with the knowledge basis, representation and proof submodule .

2.6 The Proof Submodule

It would be advisable to tackle this submodule only after the planning submodule. The student should have a choice:
- to prove the first question (if he thinks that he can), building the deductive net,
- a subproblem being supposed solved, to prove a middle step (if he thinks that he can), building the deductive net,
- more generally, a step of his plan being supposed solved, to prove another, building the deductive net,
- to gather different nets already built,
- to go back if his plan fails, while keeping the valid work: parts of valid deductive and plan nets are saved and available for building a new plan.
- to have an overview on his work as in [Gras 88]: he knows what is finished (subproblems proved) and the current plan with a representation by deductive and planning nets.

It is clear that this submodule is closely linked with the deductive organization and planning submodules.

3. Discussion

3.1 At This Moment, What Features of Such an ITS Are Workable?

- Such a system is able to comment, explain and justify its method [Nicaud & al. 88],
- The ITS has to fit with the student level [Kaltenbach & al. 89].
- A diagram construction submodule with a hight-quality interface as Cabri-géomètre [Baulac & al. 88].
- A diagram investigation submodule linked with the terms of the problem, an overview on the student's work [Gras 88].

• Possibility for the student to construct a net by himself, with a high-quality interface [Anderson & al 87].

3.2 A Short Comparison with Geometry Proof Tutor

To summarize the ideas developed in the previous paragraphs, we are in accord with some of cognitive principles of computer tutoring presented in [Anderson & al. 87]:

principle 3: Provide instruction in the problem-solving context,
principle 4: Promote an abstract understanding of the problem-solving,
principle 5: Minimize working memory load,
principle 7: Adjust the grain size of instruction with learning,
principle 8: Facilitate successive approximations to the target skill.

Nevertheless, we do not agree with:

principle 1: Represent the student as a production set,
principle 2: Communicate the goal structure underlying the problem-solving,
principle 6: Provide immediate feedback on errors.

Furthermore, the ideal student extends the proof backward from the to-be-proved-statement and forward from the givens until there is a complete proof. We have given the reasons of our divergence on these points all along this paper. Especially, concerning Geometry Proof Tutor, we have pointed out some of its educational choices:

1. The problem is given in a *predigested form*: hypothesis on one hand, conclusion on the other hand, translating natural language.

2. Three *successive* operations are asked for:
– choice of premises,
– choice of a rule,
– give the result of the rule application.

This task is too much *localized*: there is no choice (so no *risk of failing*) about the operating roles of statements. Nevertheless, the recognition of these roles is the path to come in proof processes. The student must have to *choose the order* of intermediate statements.

This task is a *mixture* of heuristic task (choose a rule) and deductive organizing task (apply a rule). Imposing a *step by step* method may be often an *obstacle* to the discovery of a solution.

The discovery of a proof is not only a *combining* of *forward* and *backward* steps. There is no possibility of *recognition* of a plan or a prototypical configuration. Unfortunately, the ideal student has to work like the novice (principle 7 does not appear).

3. The student is not *allowed* to wander too *far* off a *proof path* even if there are no errors in application of rules. We have already noticed the importance of such an exploration .

4. We do not know exactly how was made the *mathematical expertise* of this software, but the way rules are formulated is often surprising for mathematic teachers, correct solutions are not accepted (sometimes because rules are lacking) [Guin & GIA 91].

An important part of the mathematical expertise for the Macintosh version (choice of problems) was elaborated by the high school teacher who used the tutor in his classroom. Some of his *educational choices* are specific: drawings are often *particular* diagrams (for example, square for rhombus), we can find problems given with *superfluous* hypothesis: all these features are influential elements in geometry proof activity. We think that the mathematical expertise of an ITS is essential. So it has to be elaborated and discussed by a *research team..*

We have worked with Macintosh version, but the curriculum, tutoring strategy and interface functions are essentialy the same for all versions. The main differences between the Macintosh version and earlier versions on larger machines is that the earlier versions allowed backward inferences, while the Macintosh version does not. We think that Geometry Tutor would be an appropriate tool to check a geometry proof (although it requires often too many justifications), but not to help the student to discover the solution (for more details see [Guin & GIA 91]).

4. Conclusion

The elaboration of an ITS requires a detailed cognitive and didactical analysis which should be discussed by many researchers in education and experimented in classrooms.

Even if its complete realization is not already possible, it is perhaps more interesting to elaborate only some parts of it which are in accordance with the previous analysis instead of providing software not well adapted to an efficient help to learn.

5. Update (January 1993)

Franklin Boyle was expected to participate in this workshop, unfortunately he did not come to Grenoble. Nevertheless, this paper was sent with more details and comments (cf. [Guin & GIA 91]) to the authors of Geometry Tutor on November 89, but remained unanswered. However, we notice, with pleasure, a real evolution in the latest papers presented [Koedinger & al. 90; Anderson 92]:

"We found the mode of attack by human experts was distinctly different from that of Geometry Proof Tutor. It seemed important to be able to characterize this expertise both as a goal in and of itself, and for pedagogical purposes ..."

"We have found that geometry experts skip steps in developing proof plans ..."

Taking into account these points, K.R. Koedinger and J.R. Anderson have developed a theory of *perceptual chunking* based on the use of diagram by experts, including the process of recognizing configurations. They have realized a computer simulation called the *Diagram Configuration* model (DC model). "The core idea of DC model is that knowledge of skilled geometry problem solvers is organized around certain prototypical geometric figures they call *diagram configurations...* these solvers think at a *larger grain size* than the grain size at which Geometry Tutor works (step-by-step) ... A tutor based on DC configurations schemas could give instruction at the larger grain size characteristic of skilled performance, and better aid students in reaching this level of skill". We agree about this new process, nevertheless, we notice that diagrams configurations schemas seem to be clearly different from our prototypical configurations (cf. [Guin & GIA 93]).

Another cause of this evolution is "difficulties when tutors, which had some success in the laboratory, *entered* the classroom of a teacher who had a different image of what should be taught.

That same teacher had great success with Geometry Tutor *but had played a larger role* in fashioning it and found it more congenial to what he wants to teach...The major shift in our development philosophy is to focus on *educator's conception* of the skill rather than our own... We are now working with educators and teachers trying to identify from their conception of what should be taught... the educators serve as experts and we try to codify their expertise... the result is something the teachers are happy and which is viable in the classroom... the severe constraint in our case is that the expertise be modelled in a *human-like way* so that it can serve as a target of expertise".

This process is a promising way but seems to be in contradiction to: "DC Model can be taken as a *new method* for doing geometry proofs that can be explicitly taught in the classroom" [Anderson 92]: If we want to avoid conflicts with teachers when tutors enter the classroom, the ideal processes seem to us to be radically *opposite: from a didactical analysis*, based on various experimentations in classrooms (not with very few teachers), but also on an epistemological research, then the expertise can be extracted and made explicit, afterwards this expertise will be codified [Guin 91]. This is Cabri's development philosophy which requires a real interdiciplinary *research team* including researchers in mathematical education, computer and cognitive science and teachers.

References

Anderson J.R., Boyle C.F., Farell R., Reiser B.J. (1987), Cognitive principles in the design of computer tutors. In: P. Morris (ed.) Modelling Cognition, pp. 93-133, John Wiley & Sons.

Anderson J.R. (1992), Intelligent tutoring and high school mathematics, Proceedings of ITS'92, Montreal, LNCS 608, pp. 1-10. Berlin: Springer-Verlag.

Baulac Y., Bellemain F., Laborde J.M (1988), Cabri-géomètre un cahier de brouillon pour l'apprentissage de la géométrie, Editeur Université J. Fourier, Grenoble.

Duval R. (1991), Structure du raisonnement déductif et apprentissage de la démonstration, Educational Studies, 22/3, 233-261.

Gras R. (1988), Aide logicielle aux problèmes de démonstration géométrique dans l'enseignement secondaire, Petit X n°17, pp. 65-83, IREM de Grenoble.

Guin D. (1991), Nécessité d'une spécification didactique des environnements informatiques d'apprentissage, Actes des 2 èmes journées E.I.A.O de Cachan, pp. 253-260, ENS Cachan.

Guin D., GIA (1989), Réflexions sur les logiciels d'aide à la démonstration, Annales de Didactique et de Sciences Cognitives, vol 2, pp. 89-109, IREM de Strasbourg.

Guin D., GIA (1991), Modélisation de la démonstration géométrique dans Geometry Tutor, Annales de Didactique et de Sciences Cognitives, vol 4, pp. 5-40, IREM de Strasbourg.

Guin D., GIA (1993), Eléments d'expertise dans l'activité de démonstration géométrique, to appear in Annales de Didactique et de Sciences Cognitives, IREM de Strasbourg.

Kaltenbach M., Frasson C. (1989), Dynaboard: user animated display of deductive proofs in mathematics, Int J. Man-Machine Studies, 30.

Koedinger K. R., Anderson J.R. (1990), Abstract planning and perceptual chunks: elements of expertise in geometry, Cognitive Science, 14, 511-550.

Nicaud J.-F., Vivet M. (1988), Les tuteurs intelligents: réalisations et tendances de recherches, T.S.I. vol 7 n°1, pp. 21-44, AFCET-Bordas.

Modelling Geometrical Knowledge: The Case of the Student

Celia Hoyles

Department of Mathematics, Statistics and Computing, Institute of Education
University of London, 20 Bedford Way, London WC1H OAL, U.K.
E-mail: choyles@ioe.ac.uk

1. Issues in School Geometry

Geometry provides a domain in which to study and operationalise deductive methods and at the same time a means by which space can be explored inductively. These opportunities arise from two characteristics of geometry, namely its logical structure and its potential for modelling the real world. The tension inherent in endeavouring to preserve a balance between these twin features is evident in the debates over many decades about the place of geometry in the school curriculum. A report on the teaching of geometry in schools in the UK in the 1960s suggested that "neither the subject matter to which attention is invited nor the operation to which the name of proof is given should retain a uniform character throughout the school age" (Mathematical Association 1963, p. 7).

Three pervasive problems in the teaching and learning of school geometry are manifest, both in the past and the present: the separation of inductive generalisations based on an intuitive mathematization of space from formal definitions and proofs; the algebraisation of geometry and the suppression of geometric thinking; and the dominance of perception over geometrical argument. I will consider each of these briefly.

In the early part of this century, geometry was defended in the school curriculum since the "discipline in geometrical argument is a training in general accuracy" (M.A., p. 8). This position became untenable as psychologists came to agree that "it is not possible to improve a mental physique by exercising a particular logical muscle" (M.A., p. 8). Later the debate became informed by research into the teaching and learning of geometry which indicated that students found it hard to either follow or to construct chains of deductive reasoning in the context of geometry (Freudenthal 1973). The question is whether these problems arise from the obscurity of abstract argument per se or with its application to the Euclidean plane or space – that is with process or with content. Rather than trying to unpick this dilemma, Freudenthal suggests that the failure of geometry could be traced to the way geometry was taught in school: "deductivity was not taught as reinvention, as Socrates did, but [that it] was imposed on the learner" (Freudenthal 1973, p. 402). This argument resonates with a much earlier statement from Poincaré which juxtaposed logic with intuition: "it is by logic one demonstrates, by intuition that one invents" (Poincaré 1913, p. 216).

Certainly, in the past, in the U.K. at least, we find a geometry curriculum dominated by axioms, definitions, theorems and corollaries – exercises in 'pure' deductive reasoning. In such curricula, we can discern little attempt to 'connect' the mathematical objects at the heart of the proofs with students' spontaneous intuitions about their visual world, or even with some of the geometrical constructions undertaken in earlier years (see Schoenfeld 1985 for a fascinating account of how students fail to invoke proof as an explanation of a geometrical construction). Only limited opportunities were available for children to engage in any inductive thinking from which they could build deductive competences.

As Freudenthal has argued, geometry is a field in which both inductive and deductive learning should feed on each other dialectically while students shift between practical applications and abstract representations. "Geometry is one of the best opportunities which exists to learn how to mathematize reality" (Freudenthal 1973, p. 407). We should find ways to exploit this propensity without losing sight of the axiomatic side of the coin. Yet all too often we find that a balance is not struck between the two characteristics of geometry – one or other tends to find its way to centre stage. Taking the situation in the U.K. as a case in point, in the 1950s, geometry was taught as a formal system; now, in contrast, the deductive side of geometry has all but disappeared – students name shapes and explore some of their properties with no glimpse of the processes of proving or proof.

This gap between inductive generalization and deductive reasoning means, on the one hand, that students learn definitions of geometrical objects and use them within proofs without associating with them any geometrical meaning and, on the other, students explore and search for patterns without developing a sense of mathematical necessity, a feeling for the need to justify or to prove. Either way there are significant consequences. First there may be problems for the proof process. Definition forms part of a deductive chain but also is a vehicle for explanation. Yet, as Freudenthal (1973) has pointed out: "How can you define a thing before you know what you have to define?" (p. 417). But, is it possible to arrive at a definition acceptable to the mathematics community by a process of induction? Second, it is likely that bugs in students' understanding of geometrical concepts may pass unnoticed – a spontaneous (possibly incorrect) conception can stand alongside a formal definition uncontaminated by it and unchanged by its use! Consider the range of definitions of angle that can be found in school texts: "the union of two rays with a common vertex"; "an ordered pair of lines"; "the intersection of two half planes" (taken from Willson 1977, p. 59). With what intuitions of angle would these statements resonate?

A second problem in school geometry can be traced back to its algebraisation. Freudenthal (1973) has suggested that a consideration of the history of mathematics reveals how geometry was unable "to compete with the great fertility of algebra and analysis, the more it was neglected and the more its weakness became evident the more people were inclined to rely on the so called analytic geometry" (p. 420). So geometry became unfashionable amongst research mathematicians: "Geometry is dead as an autonomous branch (of mathematics); it is no more than the study of particularly interesting algebraic–topological structures" (Davis and Hersh 1981, p. 396). Although the relationship between scholarly mathematics and school mathematics is by no means straightforward, it would seem reasonable to conjecture that this view of geometry would have filtered into discussions of the school curriculum and affected geometry's position there.

Any move to ease out the visual in favour of the analytic in schools would have gathered support from difficulties associated with teaching geometry, the absence of easily definable skills and procedures 'to be transmitted from teacher to student'. In contrast to much of school arithmetic and algebra, geometrical problems tend not to fit into neat categories with routinised solution procedures, a certain amount of creative thinking is required – something to be applauded by mathematics educators but not necessarily by those concerned with the management and assessment of the school curriculum. The algebraisation of school geometry is manifested in different ways: questions ostensibly about angles are converted to exercises in finding the unknown in a linear equation, transformations are reduced to the manipulation of matrices.

One consequence of the algebraisation of school geometry is that calculation and algebraic manipulation become the focus of activity to the neglect of visual reasoning and the mobilisation of geometrical skills. For example, transformation geometry is frequently reduced to exercises in multiplying matrices rather than appreciating the variants and invariants of reflections and rotations. Algebra has the status and the potential to make problems 'handle-cranking' – so geometry is squeezed out. It is an interesting perhaps ironic fact that most of the investigations which form a major part of the present U.K. mathematics curriculum have a geometrical context – yet this context is largely ignored in the solution process. Students become centrated on finding and generalising number patterns without reference to the geometrical implications. This tendency is illustrated in some extracts of students' work shown in Figure 1. The problem was to investigate the relationship between the dimensions of a trapezium and the number of triangles it contains when drawn on isometric paper – a problem with evident geometrical content. Yet note the progression from counting to pattern-spotting to the neglect of any geometrical thinking or justification.

One of the main difficulties in learning geometry is that students have to come to appreciate that mathematical invariance is not the same as psychological invariance – images (usually on paper) which 'embody' the same mathematics may look different yet must be regarded as the same (see the discussion of the prototype phenomenon in (Hershkowitz 1989) and the litany of student misconceptions summarised in (Clements 1992)). For example, children are able to identify a right-angled triangle if the sides of the right-angle are horizontal and vertical but not if the orientation of the triangle is changed; an isosceles triangle always stands on its base; a parallelogram always has one pair of horizontal sides and tends to slope to the right. The problem for students is how to separate the critical attributes of the geometrical object, its defining characteristics, from those that are simply artefacts of the visual display; that is distinguish what Laborde (1994) defines as the figure from the drawing: "*Drawing* refers to the material entity while *figure* refers to the theoretical object" (her emphasis). I maintain this distinction in the remainder of this chapter.

The problem that we were given was to see how many triangles there are in a trapezium.

A trapezium which contains
16 inner triangles (right)
The dimensions are :-
top length 3 units
bottom length 5 units
slant length 2 units
For 12 inner triangles:-
top length 2 units
bottom length 4 units
slant length 2 units
For 32 inner triangles:-
top length 2 units
bottom length 6 units
slant length 4 units

Below shows the results of a quick conversion table. If you have a trapezium with a slant of 1 and a top of 1 you look on the table and the answer is 3

	1	2	3	4	4
1	3	8	15	24	35
2	5	12	21	32	45
3	7	16	27	40	55
4	9	20	33	48	65
5	11	24	39	56	75
gap	2	4	6	8	10

There is a gap of 2 between each answer

Base	Slant	Top	Total
2	1	1	3
4	2	2	12
6	3	3	27
8	4	4	48
10	5	5	75

So the formula is: the top + the bottom × the slant: $(t + b) \times s$

Evaluation
I found this investigation fairly easy and there were a lot of patterns and formulas to find but once our group got underway there was no stopping us.

Figure 1. How many triangles in a trapezium?
(adapted from some students' texts collected by Candia Morgan)

2. The Role of the Computer

In the previous section, I have reflected upon three problematic issues that need to be faced in any consideration of school geometry[1]. Many claim that the dynamic graphical facilities of computers have the potential to enhance geometrical understandings, to revolutionise the school geometry curriculum. I suggest that we can be more precise about the place of the computer in geometry by considering its role in relation to each one of the issues described earlier, namely in:

• Bridging the gap between inductive and deductive thinking.
• Bridging the gap between the analytic and the visual.
• Bridging the gap between the drawing and the figure

Conceptualising the computer as a a 'gap-closing' medium is not a new idea; in fact Papert (1980) suggested that the potential for computers was to "bridge the gap between formal knowledge and intuitive understanding" (p. 145). However breaking down his general claim into more specific points, generates ideas for software design and classroom activities which together could harness the computer's power for the particular purposes of learning geometry. Can we encourage students to model their geometrical ideas using software tools, to explore inductively the properties and limitations of their constructions, to reflect upon and discuss their strategies and to make the structure of a figure and the relationships between figures explicit by means of a computer language? *Can the computer interaction provide the student and teacher with a set of three two-way mirrors – to and from induction/deduction, visual/analytic, drawing/figure?*

Different geometry software is now available for use in schools each involving the manipulation of graphical objects on the computer screen by some means or other; one set is based around the programming language Logo now augmented by microworlds providing software tools specifically for exploring Euclidean geometry; the other comprises application software designed to construct and manipulate Euclidean objects, the Geometric Supposer, the Geometry Inventor, CABRI Géomètre, and the Geometer Sketchpad to name but a few. Superficially the software in both these sets might seem similar but surface similarities belie fundamental divergence in purpose and culture – in the type of actions students are expected to perform, the methods by which the geometrical objects are constructed and manipulated, the way students are supposed to justify their conclusions.

Logo is probably the most researched software in use in education. In the next section of this chapter I will look back over the decades of research into LogoGeometry in order to point to the gap-closing potential of computers, the problems that might arise and finally to draw some implications for the theme of this book, modelling student knowledge in geometry. In focussing on Logo work, let me stress that my intention is to look *through Logo* to the broader debate

---

[1]For a more detailed and comprehensive review of research in geometry and spatial reasoning, its successes and failures, see (Clements 1992).

about teaching school geometry with computers. The aim is that the reader recontextualises the stories across software rather than sees them as fuel in a technocentric debate around 'which software is best'.

3. The Potential of LogoGeometry

In the turtle graphics subset of Logo, the primitives of the language provide tools to draw and to measure lengths and angle. Lengths are measured in turtle steps and angles in terms of rotation or turtle turns. Geometrical objects are constructed and modified in direct mode by turtle commands or at a higher level of abstraction through procedures which are symbolic and formalised. Of relevance to this paper is *to view a procedure on the one hand as a summary of actions performed while building up a screen object, a drawing, while on the other, and simultaneously, as a representation of the structure and relationships of the geometrical figure.* In Logo, construction can be negotiated in action and children can build up shapes interactively, modifying their productions as a result of the visual clues from the turtle drawing. This is often called planning-in-action, a process contrasting with the style demanded by other geometrical software where construction requires explicit prior description. Procedures written in Logo can themselves serve as new primitives in the language and as objects to be explored in their own right. The way a procedure is built offers a formal, precise description of the structures and relationships of the geometric object, albeit in a form that is specific to turtle geometry and to the Logo language[2].

There is now a considerable corpus of research into the potential of Logo for learning geometry. This research can be grouped under different headings: work which relates specifically to children's understanding of angle (e.g., Noss 1987); investigation of the interrelationship between perceptual schema and the mobilising of geometric knowledge (e.g., Hillel & Kieran 1988); and analyses of the relationship between geometric understandings developed in a Logo context with the meanings operationalised in other contexts such as paper and pencil (e.g., Clements & Battista 1990). This research is summarised in (Clements 1992), and in the chapters of the edited volume (Hoyles & Noss 1992), particularly those by Edwards, Kynigos and Leron and Zazkis. It is probably fair to say that "there is evidence in support of the hypothesis that Logo experiences can help elementary to middle school students become cognizant of their mathematical intuitions and facilitate the transition from visual to descriptive/ analytic geometric thinking in the domains of shapes, symmetry, and motions" (Clements & Battista 1990).

So what might be the features of the turtle graphics environment that might explain these results? I suggest the following.

---

2 Numerous texts are available about Logo and turtle geometry, the best known being (Papert 1980, Abelson and diSessa 1980).

- Students begin work with Logo in *direct mode*; they identify with the turtle and build up shapes by immediate interaction, shapes whose properties can be generalised and tested out *inductively*. The graphical objects cannot be manipulated directly, but students can *directly experience* their construction.
- Logo has the potential to *link visual and symbolic representations* in a relatively natural way – there is a graphical output built up by means of the symbolic code, the Logo program.
- Logo is an extensible language, so teachers and students can build their own Logo tools and design their own microworlds for investigations of geometrical ideas. There can be a *variety of approach and style* with different mixes of on/off computer activity.
- The syntax of Logo programs can be constructed so as to be mathematically expressive; the structure of the *geometric object can be seen through the window of the Logo code, the figure seen through the drawing.*
- Finally, and in contrast to most other geometry software, the goal of Logo work, at least from the student's point of view, is not frequently mathematics – *the mathematics is hidden whilst remaining crucial as a tool to achieve a student's goal.* Motivation is triggered by the student's project, the desire to build and create something for oneself.

I will now try to illustrate some of these points by reference to a hypothetical example stimulated by some of our work in London (see Hoyles & Noss 1989).

Janet sets out to construct a street of 3-dimensional apartment blocks. Since there are so many parallelograms in the drawing she decides to construct a procedure for a parallelogram which can be used for the windows, roofs, doors, etc. Her first attempt is a fixed shape, PARALELLOGRAM1:

```
TO PARALELLOGRAM1
FD 50 RT 30
FD 100 RT 150
FD 50 RT 30
FD 100 RT 150
END
```

This procedure can be built in direct mode and then subsequently made into a procedure or constructed directly in the editor. Even though the parallelogram drawing might at first have been built by exploiting perceptual cues, 'seeing that it looks just about alright', the procedural representation captures a pattern of commands by means of which Janet can discriminate the significant features of the geometrical figure: she might notice, for example, that the alternate inputs to FD are the same – the Logo language way of depicting that the opposite sides of a parallelogram are equal; or that alternate turtle turns are the same – the Logo equivalent (indirectly!) that opposite angles of a parallelogram are equal. Janet needs parallelograms of different sizes in her picture. She can use her fixed procedure as a template to generalise the parallelogram properties:

```
TO PARALLELOGRAM2 :S1 :S2 :A1 :A2
FD :S1  RT :A1
FD :S2  RT :A2
```

```
FD :S1  RT :A1
FD :S2  RT :A2
END
```

Evidently this is not a completely general procedure for a parallelogram – PARALLELOGRAM2 will only output a parallelogram if the two angular inputs A1 and A2 happen to add up to 180. But Janet plays around with different values for the four inputs and is led to this conjecture in an inductive way formalising it as follows:

```
TO PARALLELOGRAM3 :S1 :S2 :A
FD :S1  RT :A
FD :S2  RT 180 - :A
FD :S1  RT :A
FD :S2  RT 180 - :A
END
```

By playing with PARALLELOGRAM3 and using it in her apartment project, Janet becomes confident that she has correctly identified and made explicit the pertinent features of a parallelogram – wherever she places PARALLELO-GRAM3, whatever its orientation and regardless of the inputs chosen, the graphical outcome appears to have its opposite sides parallel. The program captures what Janet visualises as the geometrical object and at the same time provides her with a means to reflect upon the structure of the geometric object, to manipulate it and to generalise the crucial features in a step by step manner – the three gap-closing potentialities of computer work in geometry. The symbolic program facilitates reflection – on its own structure and that of the drawing – and thus helps Janet see what can be varied and what cannot. Thus any specific screen object is not in fact seen as such – but rather as a generic example of a whole class of objects satisfying the same structure.

We would expect that having experienced this constructive cycle, Janet would formulate her own LogoGeometric definitions of a parallelogram: 'any object which can be drawn by the procedure PARALLELOGRAM3', or 'any object which can be drawn by the procedure PARALLELOGRAM2 provided the third and fourth inputs add up to 180'. Richard Noss and I have termed such descriptions, *situated abstractions,* (see Noss & Hoyles 1992; Hoyles 1994; Noss, in press). A situated abstraction is a student articulation of a general mathematical structure or relationship which is *constructed by the student through an inductive process* and which is *expressed by the student in terms of the medium of construction.* In this case the situated abstraction is expressed through Logo. The way a student is able to construct her own situated abstraction of a geometrical idea points to the mechanisms by which computer use can bridge the three gaps in geometry learning described earlier – situated abstractions are arrived at inductively by experimentation yet are formalised; they derive their semantic sense from visual as well as symbolic cues and by their articulation exhibit an appreciation of the general figure underpinning any particular instantiation on the screen.

Children inevitably do not always learn just what we expect them to learn and there are of course obstacles to understanding. Many have been described elsewhere (see for example Noss & Hoyles 1992) and some can be illustrated by reference to the apartment project. Do children notice geometrical structures when their attention is focussed on the goal of their project? What are the meanings that children have of the input to RT? Do they 'see' this as the exterior angle of a parallelogram? What connections are made (if any) between the situated abstractions articulated in a Logo environment and the declarative definition of a parallelogram presented in a school textbook? These are open questions some of which will be considered in the next section.

4. Learning from the LogoGeometry Experience

I will look back over the work we have undertaken in the turtle graphics subset of Logo to see what we can glean from our experiences which might throw light on these questions and on the role of the computer in learning geometry. Given the title of this book, it is important to make clear that at no time did anybody in the LogoMathematics community – as far as I am aware – set out to try to model student knowledge, to set up an 'intelligent Logo tutor'. This I contend is no coincidence. LogoMathematics came out of a very different culture from that of computer tutoring – a culture which set great store on student autonomy and students constructing their own mathematics whereas much of the ITS community emerged from a culture which eschewed constructivist tenets implicitly or explicitly.

It is clear that the potential of LogoGeometry has *not necessarily* been fulfilled. To take just one illustration, although Logo is extensible, it is rare indeed to see teachers, let alone students, design their own microworlds. But we have a problem here which is hard to unravel – is this because of difficulties in the language or the constraints of the culture of school mathematics? More generally, are obstacles bound up with a particular software or with issues around teaching and learning school geometry? Logo activities are not immune from the influence of society, school and classroom climate (see Hoyles 1994) and this is true of computer use in general. We must not neglect to search for reasons beyond the pros and cons of any specific software.

Our experience with LogoGeometry points to problems of incorporating geometrical software in school mathematics which I shall describe through a set of stories from our research in London – stories which I hope will capture the general in the particular![3]. The setting is Logo but the issues are general. The intention is that each story will generate inductively a set of issues for the readers of this book and the software developers of the future.

---

[3]Other researchers have come up with similar findings so the observations are reasonabl robust.

Story 1:   Children working with computers become centrated on the screen product at the expense of reflection upon its construction

In the Logo Mathematics project (Hoyles & Sutherland 1989), we observed pairs of children mechanically copying programs – from each other or from the teacher. One pair in particular 'discovered' a set of commands which produced an impressive outcome – REPEAT :X [ANYPROCEDURE RT :Y]. They happily inserted different inputs into this pattern for X, Y and ANYPRO-CEDURE and watched the pretty graphical display. They were not interested in how their display was constructed, resisted any attempts at explanation and refused to take part in any tasks which might have involved them in relating process to outcome.

Story 2:   Students do not mobilise geometric understandings in the computer context

In the Logo Mathematics project, we followed John and Panos's developing understanding of 360 degrees over a period of seven months. Before starting the Logo activity Panos and John had successfully completed work from a mathematics booklet which 'covered' 360 degrees as the angle around a point. However, when attempting to draw a complete Logo circle they consistently adopted an experimental trial-and-error approach – failing to 'apply' their knowledge of 360 degrees. Their work was not unreflective and they constructed a range of situated abstractions about how to proceed. From a starting point of counting – 16 REPEATs needs a turn of 22.5 – they worked out a proportional strategy 'To get any circle you can multiply the 16 by any number provided you divide the 22.5 by the same number'. Underlying this method was an awareness that the product had to be invariant although no connection was made between the size of the invariant and 360. The boys were centrated on relationships between numbers. Panos and John, who 'knew about' 360 degrees in the static context of their normal mathematics curriculum, did not mobilise this knowledge in LogoGeometry. This suggests that 'transfer' across settings is by no means automatic.

Story 3:   Students modify the figure 'to make it look alright' rather than debug the construction process

The third story occurred during an investigation of a ratio and proportion Logo microworld (Hoyles, Noss and Sutherland 1989). One of the computational objects which formed the basis of exploration in the computer-based activities was HOUSE, a fixed procedure drawing a closed shape from which students had to *construct* sets of proportional objects using scalar operators (see Figure 2).

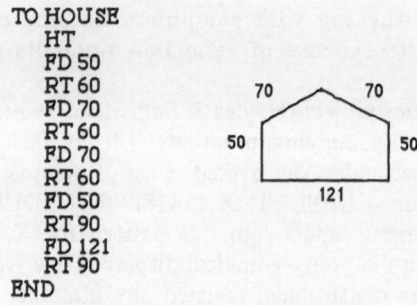

```
TO HOUSE
  HT
  FD 50
  RT 60
  FD 70
  RT 60
  FD 70
  RT 60
  FD 50
  RT 90
  FD 121
  RT 90
END
```

Figure 2. HOUSE, a fixed procedure

Activities were designed so that students would come up against visual conflict if they did not use proportional strategies for enlarging and diminishing HOUSE. It was anticipated that attention would be drawn to the necessity of multiplicative relationships since the adoption of non-multiplicative strategies produced drawings which were self-evidently not houses – such as non-closed or overlapping shapes. Figure 3 illustrates the computer feedback on the adoption of an additive strategy and the obvious mismatch between the intended and actual outcome.

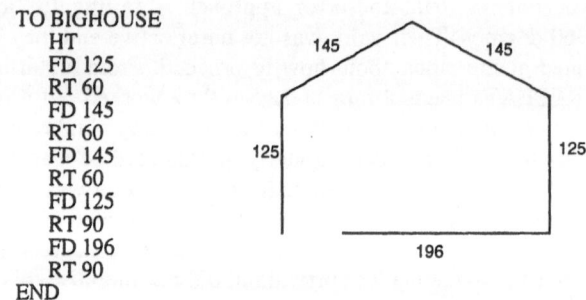

```
TO BIGHOUSE
  HT
  FD 125
  RT 60
  FD 145
  RT 60
  FD 145
  RT 60
  FD 125
  RT 90
  FD 196
  RT 90
END
```

Figure 3. The result of using an additive strategy to enlarge HOUSE

Contrary to our expectations, we noticed that many children did not experience cognitive conflict after adding and receiving the visual feedback shown above. They saw that their drawing was not a house but debugged this perceptually, simply 'closing the gap' by adding some more turtle steps to the base of the house after the figure had been drawn. These students centrated on the screen outcome, ignoring the construction process and the mathematical requirements for making a proportional product. In effect, they *debugged the drawing and not the figure as manifested in their Logo program.* The question we have to ask ourselves is why? I suggest a clue to the answer is to probe: *what are the goals of the activity for the student?*

Story 4:   Students do not appreciate how the computer tools they use constrain their behaviour

In the Logo Mathematics Project, we noticed that throughout the whole of their first year of programming in Logo, Linda and Jude restricted their input to turtle turn to multiples of 45 which were less than or equal to 180. This strategy for angle input influenced the shapes they chose to produce – squares, cubes, rectangular letters – and enabled them to cope easily with parallel lines or symmetrical slanting lines. At the beginning of their second year we suggested that they draw a regular hexagon in order to provoke them to think about turn and use a wider range of inputs. The pair immediately tried FD 40 RT 45 repeated 6 times and much to their surprise produced an incomplete octagon. They had not noticed how they had been restricting the set of numbers that they used as inputs to their turns nor reflected upon the relationship between this turn and the number of sides. After some thought and experimentation they eventually came up with a turn of 60 to produce the regular hexagon.

After finishing this hexagon we asked them to draw an octagon – something which they had done many times before using their 45-degree strategy. But they could not now see how to proceed – 45 was not their automatic choice of turn. Instead they used an inductive 'pattern-spotting' strategy unrelated to the geometric context: '6 sides for 60 degrees' so '8 sides for 80 degrees'!

Story 5:   After making inductive generalisations, students frequently fail to apply them in a new situation

Jim was asked to draw four equal squares in a line. He confidently defined the procedure for a square in the normal way:

```
TO SQUARE
REPEAT 4 [FD 50 RT 90]
END
```

When Jim came to construct his row of squares he wrote a procedure ROW which called SQUARE and then a sequence of simple turtle commands to map out the interface between the first and second squares:

```
TO ROW
SQUARE
PU
RT 90
FD 60
LT 90
END
```

Now the turtle was in the position to draw the second square but instead of using SQUARE again, Jim 'reverted' back to working in direct mode, typing in the commands:

FD 50, RT 90, FD 50, RT 90, etc.

Thus although a procedure for a square had been built by the student himself, it was not exploited to full effect – somehow or other it was not 'owned' by Jim, not part of his repertoire of tools to be applied in different problem situations.

Story 6: Students have difficulty distinguishing their own conceptual problems from problems arising from the way the software happens to work

Sally was set the task to draw five squares in a line decreasing in size. She defined a procedure, BAG, for a variable square – including the interface commands in the procedure:

```
TO BAG :SIDE
REPEAT 4 [ FD :SIDE LT 90 ]
PU
FD  :SIDE  + 10
PD
END
```

Sally then wrote the procedure, BIGBAG, for the five decreasing squares:

```
TO BIGBAG :SIDE
REPEAT 5 [ BAG :SIDE - 10]
END
```

She typed BIGBAG 60 feeling quite confident that this would work, arguing that the first square would have side 60, the next 10 units less and so on. As it happens Logo does not work this way and one square of side 50 was drawn five times on the screen.

Sally was genuinely surprised at this outcome as she had been certain that the sides of each square drawn would have been reduced sequentially by 10. What she had done indicated that she did not know the specificities of how control is passed in Logo – an interpretation supported by her debugging strategy. Sally decided that she had used the REPEAT command in the wrong way and typed:

```
TO   BIGBAG :SIDE
BAG  (:SIDE - 10)
BAG  (:SIDE - 10)
BAG  (:SIDE - 10)
BAG  (:SIDE - 10)
BAG  (:SIDE - 10)
END
```

Of course this program reflects once again how Sally saw the problem – correctly maybe but not the way Logo happens to 'see' it. Sally did not understand how values of variables are passed from a Logo procedure to subprocedures and this stood in the way of her formalisation. It suggests the need for some tuition (in the right place at the right time!) about the mechanisms of the the software – sometimes software just does not allow you to model the situation as you see it, even if this model is perfectly correct!

Story 7:   Manipulating drawings on the screen does not necessarily mean that the conceptual properties of the geometrical figure are appreciated

Nicola had built a procedure, TRI, for an equilateral triangle:

```
TO TRI
REPEAT 3 [FD 100 RT 120]
END
```

This draws a triangle with one vertical side. In her drawing for a house Nicola needed a triangle with one side horizontal for her roof. Instead of simply orientating the turtle and calling TRI, she thought she needed a new geometrical object and constructed another procedure NEWTRI with an interface turn of RT 30 as the first command:

```
TO NEWTRI
RT 30
REPEAT 3 [FD 100 RT 120]
END
```

Every time Nicola wanted an equilateral triangle oriented differently she insisted on writing a new procedure with a different name and a different first line. Obviously the structure of the REPEAT line was 'transferable' but, despite our interventions, she refused to use the first triangle procedure as an object which could draw triangles in any orientation.

Story 8:   Children do not read error messages and even if they do find them hard to interpret

Whenever Richard and Susan made a mistake, as soon as an error message came up on the screen they quickly typed return as many times as were needed to have a clear textscreen. The pair recognised that something was wrong with their program – and wanted to put it right – but did not want to leave a trace of the computer's evaluation of their efforts on the screen. They did not realise that the computer was doing more than telling them they were wrong and their behaviour meant that they did not have the chance to find out!

After some intervention – they needed to be orientated to try to make sense of feedback – the pair began to read the messages and with more experience learned both the meaning of phrases such as "I don't know how to" and more importantly what was the probable cause of the complaint (you have forgotten to put the : in front of a variable so Logo was looking for a procedure). But the software is not always helpful: some error messages are difficult to interpret, ("you don't say what to do with —"), some obscure ("too much inside parentheses") and some completely unhelpful (IF doesn't like "F as input – in response to IF FIRST :NAMES = "CELIA [....])!

Story 9:   Many children and still more teachers are nervous of machines, particularly if they seem 'out of control'

Jacky had been using Logo with her class of 11-year-olds for about a year and was growing in confidence both in her own competence and in how to organise the class to exploit Logo's potential for mathematical learning. She typed into the editor a recursive procedure PATTERN – without a STOP rule – explaining the structure as she went along. Having left the editor, Jacky typed PATTERN 40 and they all watched the complicated drawing unfold. However, fascination turned to horror as Jacky realised that she did not know how to stop the program. She randomly pressed keys, panic mounting. In desperation she turned off the monitor – at least if they did not see this uncontrollable 'monster' it would somehow go away!

Story.10:   Computers can provide scaffolding for geometrical ideas provided that these are within the student's zone of proximal development[4]

A group of 10-year-old children were asked to construct a series of regular polygons in Logo, to name the shapes and tabulate the number of sides and the turning angle in each case. They were then asked to try to find a relationship between the number of sides and the turning angle, explain why this relationship might be true in terms of the geometry of the figure, and finally, formalise it in a Logo procedure which would draw any polygon. Each of these questions proved to be difficult for different pairs of children.

One group of students saw the goal of drawing up the table as a recording device and not as a means to generalise – they simply took each polygon in turn and documented the number of sides and the turning angle. Even with help, they were unable to spot any pattern in the figures. Another pair became centrated on the numbers and found a rule connecting them, but were unable, again despite assistance, to relate this rule to the geometry of the figures, i.e., to see it in terms of number of sides of the polygon and exterior angle.

A third pair managed both these steps and explained the rule: "If you times the number of sides by the turning angle you get 360". They wrote this into a Logo procedure, TIM:

        TO TIM :SIDE
        REPEAT :SIDE [FD 10 LT :SIDE]
        END

They typed TIM 120 and much to their pleasure obtained a triangle – which happened to be drawn 40 times! Since they had certainly answered our question and built a procedure which they could make draw any regular polygon they were perfectly content with what they had done. It is noteworthy that the project was not theirs. I asked them to explain what they had done and they answered: "If you want a square, 4 repeats, the turn is 90; with a triangle, you need 3 repeats

---

[4] I refer to Vygotsky's (1978) well-known concept here.

and the turn is 120". Yet, they could not translate their verbal description into Logo code. Recontextualisation between different means of expression is not trivial. But it was more than this – the students were just not able to change the relationship $X * Y = 360$ into $Y = 360 / X$, and perhaps could not see the point of doing this anyway.

5. Modelling Student Knowledge?

Reviewing the themes of this paper, I have identified the major problematic issues in the teaching and learning of school geometry and pointed to the role of geometry software as a means to face up to these, particularly by identifying it as a way to bridge the gaps between induction/deduction, analysis/visualisation and figure/drawing. I illustrated these potentialities by reference to an example of 'successful' microworld activity where a student was actively engaged in expressing mathematical ideas in the pursuit of her own goal. I then sought to show that despite the potential of computer use in learning geometry, there remain obstacles to learning. Mathematics is not trivial to learn and concepts need to be 'attainable' given the background and experience to the student. But even if this is the case the path to understanding is littered with obstacles which I illustrated by means of ten stories. In some ways these stories highlighted the obvious: *how can students appreciate what they have yet to experience whether it be the generality of a mathematical concept or the expressive functionality of the medium? How can students 'know' where the work is to lead before they have got there?* It is also clear that *more attention needs to be paid to the process of recontextualisation – to the synthesising of situated abstractions expressed in different media.* These are general issues about learning but are magnified through the lens of the computer.

The stories also illustrate how students' investigations are moulded by the culture of school, the domination of the task in hand – 'to get the job done' at the expense of reflection on how it is done. They show how students tend to work in a local way – to react to the 'here and now' rather than draw on their prior experiences. They show how hard it is for students to acquire 'ownership' of the software and the goals of the activity. Perhaps they can also serve as a warning to software developers that the *more powerful and expressive the computer environment the more likely it is that students will bypass the mathematical agenda* – social and cultural influences will always mediate between conceptual understanding and behavioural outcome.

Given the focus of this volume, one is provoked to ponder whether one way forward would be to replace the exploratory microworld approach with a more directed tutoring system, one which attempts to model student geometrical knowledge and on the basis of this model provide guidance? Such an approach raises a host of new issues which I will raise briefly in some concluding comments.

First, using the computer as a means to model and teach geometrical knowledge casts it in a very different role from that envisaged in this paper. This new function needs to be justified in relation to the problematic issues about learning geometry raised at the beginning of this chapter. Certainly the computer's potential as a gap-closing medium is no longer tenable, since it assumes the existence of an optimal solution and implicitly denies the validity of the student's existing knowledge.

Second, the idea of modelling student knowledge gives an impression that the model is both fixed and knowable. But evidence suggests the opposite – that knowledge is contextualised and often unarticulated. I suggest that student knowledge of geometry and the situated abstractions they construct of geometrical concepts are mediated by the computer tools available, in fact by the medium in which the knowledge is elicited, whether paper and pencil, ruler and compass or computer software. This calls into question the whole notion of one model of geometrical knowledge. How is this model expressed? What will be the means of representation?

Student knowledge is also in transition. Learning takes place over time with a granularity much more delicate than that of behaviour. Knowledge is idiosyncratic, affected by the meaning of the task for the students, the expectations, goals and intentions of the students. How a student reacts in any environment will be affected by past experience, by the way the question is posed and the signals that the teacher gives as to the solution process.[5] As I have remarked in an earlier paper: "Ignoring intuition and cultural setting assumes that the student is engaged in the activity as prescribed by the teacher" (Hoyles 1990).

All this is not to say that we should not seek to be more precise as to what students might bring to school geometry; to endeavour to make predictions as to what they might do in any activity in order to optimise the help we can offer. What would be useful is to try to map the range of expected student responses to a geometrical problem and to interpret these as the images of two mappings: one with a domain of geometrical structures and relationships and the other of student conceptions of the situation. Since the construction of these maps acknowledges their situated character, the analysis of student responses will recognise how they are framed by the software tools available and by the social and affective factors which form part of any activity. The maps might provide mirrors onto how a student sees the situation but only through the lens of the medium in use, the context evoked. They will not and cannot model knowledge in any general sense.

Yet even this more limited goal begs the question as to how students react to feedback in a tutoring environment. What are the invisible messages which go hand in hand with student modelling and computer teaching? It seems to me that these modes of instruction are grounded in the notion of the expert and the novice: the expert (the computer) knows best and the novice (the student) has misconceptions which are in need of remediation. How does this view of the student configure their computer use? What does the tutoring environment mean to students? What are the expectations engendered? Will they be able to interpret

---

[5]These remarks echo some of the arguments for a constructivist approach to mathematics learning; see for example (Davis et al. 1990).

the feedback and if so, for what purpose: as a pointer for them to make mathematical sense or as a guide to tell them what to do next? Will students want to participate in this game or find ways to subvert it? [6]

Whether it is possible to model geometrical knowledge is an open question. How far we wish to pursue this goal and how we evaluate our success will depend on our theory of learning and what we see as the aims of teaching mathematics. It also depends on how we define the challenge – a search for a general model or for a set of intersecting and local maps? More fundamentally, we must face questions of a political rather than an educational nature. What is the intention behind modelling student knowledge? Is it in fact to facilitate the work of teachers or is it rather to replace them?

Acknowledgement

I would like to thank Richard Noss for his helpful comments on this paper.

References

Abelson, H., diSessa, AA (1980) Turtle Geometry: the computer as a medium for exploring mathematics. Cambridge, MA: MIT Press.

Clements, D.H., Battista, M.T. (1990) The effects of Logo on children's conceptualizations of angle and polygons. Journal for Research in Mathematics Education 21, 356-371.

Clements, D.H., Battista, M.T. (1992) Geometry and spatial reasoning. In: Grouws, D.A. (ed.) Handbook of Research on Mathematics Teaching and Learning: a project of the National Council of Teachers of Mathematics. pp. 420-464. NY: Macmillan.

Davis, P.J., Hersh, R. (1981) The Mathematical Experience. Boston: Birkhauser.

Edwards, L.D. (1992) A Logo microworld for transformation geometry. In: Hoyles, C., Noss, R. (eds.) Learning Mathematics and Logo. 127-155. Cambridge, MA: MIT Press.

Freudenthal, H. (1973) Mathematics as an Educational Task. Dordrecht: Reidel.

Hershkowitz, R. (1989) Visualization in geometry – two sides of the coin. Focus on Learning Problems in Mathematics 11, 61-76.

Hillel, J., Kieran, C. (1988) Schemas used by 12-year-olds in solving selected turtle geometry tasks. Recherches en Didactique des Mathématiques 8 (1.2), pp 61-103.

Hoyles, C., Noss, R. (1987) Children working in a structured Logo environment: from doing to understanding. Recherches en Didactique des Mathématiques 8(1.2), 131-174.

Hoyles, C., Noss, R. (1992) A pedagogy for mathematical microworlds. Educational Studies in Mathematics. 23, 31-57.

---

[6] For example, students might use the HELP facilities in an intelligent tutoring system in a strategic way which gets them to the solution without having to use any geometrical knowledge.

Hoyles, C., Noss, R. (eds.) (1992) Learning Logo and Mathematics. Cambridge, MA: MIT Press.

Hoyles, C., Sutherland, R. (1989) Logo Mathematics in the Classroom. London: Routledge.

Hoyles, C. (1990) 'Neglected voices: pupils' mathematics and the National Curriculum' In: Dowling, P., Noss, R. (eds.) Mathematics versus the National Curriculum. Basingstoke: Falmer.

Hoyles, C. (1994) Microworlds/Schoolworlds: the transformation of an innovation. In: Keitel, C. & Ruthven, K. (eds.) Learning from Computers: mathematics education and technology. NATO ASI Series F, Vol. 121. Berlin: Springer-Verlag.

Hoyles, C., Noss, R., Sutherland, R. (1991) The Microworlds Project, 1986-89, Final Report. London: Department of Mathematics, Statistics and Computing, Institute of Education, University of London.

Kelly, G.N., Kelly, J.T., Miller, R. B. (1986-87) Working with Logo: do 5th and 6th graders develop a basic understanding of angles and distances? Journal of Computers in Mathematics and Science Teaching 6, 23-27.

Kieran, C. (1986) Turns and angles — what develops in Logo? Proceedings of the Eighth International Conference for the Psychology of Mathematics Education North American Group, pp. 169-177.

Kynigos, C. (1992) The turtle metaphor as a tool for children's geometry. In: Hoyles, C., Noss, R. (eds) Learning Mathematics and Logo, pp. 97-126. Cambridge, MA: MIT Press.

Laborde, C. (1994) The computer as part of the learning environment: the case of geometry. In: Keitel, C. & Ruthven, K. (eds.) Learning from computers: mathematics education and technology. NATO ASI Series F, Vo. 121. Berlin: Springer-Verlag.

Leron, U., Zazkis, R. (1992) Of geometry, turtles and groups. In: Hoyles, C., Noss, R. (eds) Learning Mathematics and Logo, pp. 319-3 52. Cambridge, MA: MIT Press.

The Mathematical Association (1963) The Teaching of Geometry in Schools. 4th ed. London: G.Bell & Sons.

Noss, R. (1987) Children's learning of geometrical concepts through Logo. Journal of Research in Mathematics Education 18 (5), pp. 343-362.

Noss, R. (in press) Meaning mathematically with computers. In: Bryant, P., Nunes, T. (eds.) Cambridge University Press.

Noss, R., Hoyles, C. (1992) Looking back and looking forward. In: Hoyles, C., Noss, R (eds) Learning Mathematics and Logo, pp. 431-468. Cambridge, MA: MIT Press.

Papert, S (1980) Mindstorms: children, computers, and powerful ideas. Brighton: Harvester.

Poincaré, H (1913) The value of science. In: The Foundations of Science (trans. Halstead, G.B.) New York: Science Press.

Schoenfeld, A.H. (1985) Mathematical Problem Solving. London: Academic Press.

Willson, W.W. (1977) The Mathematics Curriculum: Geometry. Glasgow: Blackie for the Schools Council.

Intelligent Microworlds and Learning Environments

Jean-Marie Laborde

LSD2-IMAG, BP 53, 38041 Grenoble cedex 9, France
E-mail: Jean-marie.Laborde@imag.fr

Abstract. Recent developments in computerised learning environments have given the concept of microworld a privileged status. In this paper we look at the conception of learning environments based on microworlds which manage knowledge explicitly.

In recent times, traditional computer assisted teaching has been going through a crisis, due essentially to its inability to fulfil its promises. At the same time the success (indisputable on certain points at least) of Seymour Papert with the promotion of LOGO has contributed a new source of reflection on the (inevitable) development of computers in the area of education and training. In this paper we propose a general framework for the apprehension of the notion of microworld. In particular it will permit the identification of a certain number of problems in the context of its use as a basis for an intelligent tutorial system in the orientation of the principles of *guided discovery learning,* which have been developed in particular by the schools of Lancaster or Leeds in Great Britain around the ideas of John Self and [Burton & Brown 79]). The Cabri-géomètre program[1] developed as part of an IMAG project [Baulac, Bellemain & JM

---

[1] The project for a Computerized Interactive Sketchpad for Geometry first saw the light in 1986, in the orientation of the CABRI project which was initially centered on the theory of graphs [Habib & Laborde 1984] .

The initial specifications foresaw "the creation of an interactive system for the processing of geometrical figures; [...] The primitives of the drawing of the figures will be available from running menus [...], it will be possible to modify the whole of a figure by only changing the characteristics of the basic elements which determine it, [...] as each figure will be structurally linked, in logical sequence, to primitives put to work in its construction, the user will be able to indicate the required calculations by the running menus or [direct] introduction of arithmetical formulae." [Laborde 1985].

The user can distinguish the tools put at his disposal by the software as an aid to help him solve the problems: he can use them to explore the properties and the exceptions envisaged by or passed over by classical statements. Indeed, he can rapidly multiply the exceptions, envisage or on the contrary eliminate certain special cases (by choosing for example the figure which appears to be "the most general", visualise the common properties which a set of figures may possess (highlighting of invariance representative of certain properties), or follow the movements of one or more elements according to the movements of one of the basic elements of the figure.

Laborde 88], [JM Laborde & Trilling 89] will act as a special example of the confrontation (validation) of the points of view developed below.

1. Of Microworlds

The concept of microworld has no clearly established origin, its first appearance seems to have been in an internal report of the AI laboratory of MIT, by Minsky and Papert [Minsky & Papert 72]. The primitive idea is obviously related to the term of microcosm. It was Papert's book *Mindstorms: Children, Computers, and Powerful Ideas* which made the expression public, and at the same time, with LOGO, brought out a new way of using the computer in class ("computers no longer programme the child, but the child programmes the computer").

The origins of LOGO themselves date back to the conception of a sort of minimal set of orders in natural language for commanding a robot [Minsky & Papert 69]. The initial project developed to make LOGO a programming language including recursiveness (contrary to BASIC), and whose central use is to pilot a turtle, which can leave a trail when it moves about. The team which worked around Papert thought that such a programming environment could bring about in children the urge to explore the numerous possibilities of the use of computer programmes: the user's interest is sustained by, among other things, the immediate effects on the computer screen which the language allows to be obtained relatively easily.

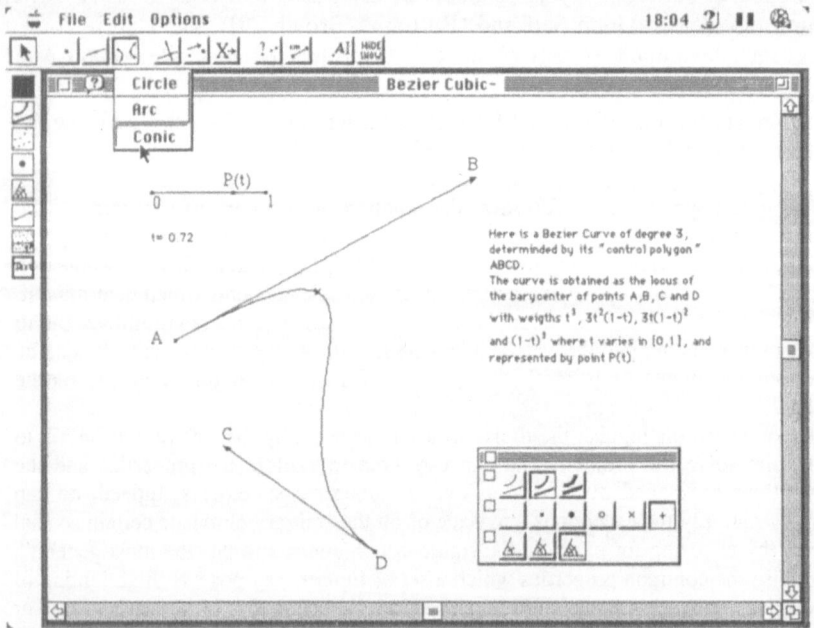

Figure 1. A general example

On this occasion the expression "microworld" began to spread, to designate certain environments in which the creative activity of the user is supposed to be particularly stimulated. Schematically, the whole of the LOGO philosophy is based on this idea, originally from Piaget, in particular that popular among its numerous protagonists in school.

There is no patented definition of the concept of microworld. The fathers of the concept, Minsky, Papert, and from a distance, Winograd and Lawler, have never really provided one. It is however agreed [Thompson 87, Laborde 89] to estimate that a microworld is made up of:

• an environment of objects and relations,

• a set of operators able to operate on these objects by creating new ones which present certain new relations.

One should probably [JM Laborde 89] add:

• a more or less clear relationship with the concept of direct manipulation.

The first remark to be made is that the definition could be reduced and one could consider an environment as simple operators, as it is always possible to represent a relation in the form of an operator. The second remark will be taken up again later and emphasises that direct manipulation is not a minor point: the first LOGO turtles were cybernetic turtles, which were manipulated directly by the use of buttons on a remote control; it would seem almost obvious that it was only because of unavailability at the time of corresponding concepts (which were already in preparation but on the other side of the continent, on the West Coast of the United States), that the classical LOGO did not use direct manipulation. In fact one can only be surprised at the relatively poor quality of implementation given to LOGO, even on fairly expensive machines such as the PDP-11s of its first available implementations[2].

We would like to give a few examples of microworlds in order to illustrate our subject:

• the SHRDLU language [Winograd 72] and its manipulation of blocks (at the origin of the concept);

• LOGO, presented above;

• any programming environment which allows new objects to be produced from initial elements. A spread-sheet type software can be considered from this angle;

• QUEST, an intelligent tutor for training in the break-down of electric circuits [White & Frederiksen 85];

• LOGO in its turtle-board version in which the manipulation primitives of the turtle apply, to a certain extent, to a turtle symbolised by one of the pupils in the class, and, in a relatively similar way,

• to a remote-controlled turtle model which is technologically refined to a greater or lesser extent;

---

[2] In many implementations of LOGO the user, who makes his turtle disappear towards the top ends up seeing it reappear from the bottom of the screen, still moving in the same direction. Many reasons for this behavior can be imagined which, however, means that the geometry of the turtle is more that of the tore than of the plane.

- Cabri-géomètre, a software which allows the creation on screen of any elementary geometrical figure and allows their parameters to be modified at will (see note 1 and [Baulac 90]);
- Planets from the Maths Dept of Cornell University (Ithaca USA), a software which allows a system of sun(s) and planets to be simulated, and their trajectories to be observed according to the different possible settings of parameters which represent "initial conditions"[3] ;
- an air-cushion table which allows the movements of material bodies not subjected to frictional force to be simulated in two dimensions;
- Steamer [Stevens & al. 83, Forbus 84, ...] often considered to be simply an "interactive simulation" on the running of the boilers and machines of a ship, whereas its direct manipulation microworld aspect is obvious;
- any word processing programme (is a word processing programme fundamentally different from a programming environment?).

Some microworlds use mathematics, others physics. A team from Geneva [Boder 90], using Hypercard, have conceived an environment (Ramos) in which the user is asked to organise a defence system to plead the cause of a child from a run-down suburb of a South American town (the child is accused of drug trafficking); numerous parameters can be brought into play such as family, social, economic, political parameters and the success of the plan can be tested; this is also a microworld.

We intend to place these relatively disparate situations in a single setting which will allow us to identify some of the problems posed by

- the use of a microworld for teaching aims,
- the status of microworld as a basis for the future development of computerised tutorial systems,
- the question of the knowledge used in the use of a microworld, making it *more or less intelligent.*

2. The Real World–Model–Microworld Trilogy

Each of the examples presented brings into play three components:
- **a system of objects which represents a frequently complex reality.** This may, for example, be a cosmic system, a thermal power station, our immediate "geometrical" space;
- **a theory which modelises the preceding system,** a sort of formal system, which stems more from the area of ideas. This is Newtonian mechanics, control theory, axiomatic geometry;
- **a materialisation of the concepts of the preceding theory** (objects and relations) accompanied by operators, and which makes up the corresponding microworld. This is the Planets software, the Steamer environment, LOGO or Cabri-géomètre.

---

[3] One could also mention along the same lines more recent creations such as the Gravitation software by Jeff Rommereide (Laurel Springs, NJ, USA).

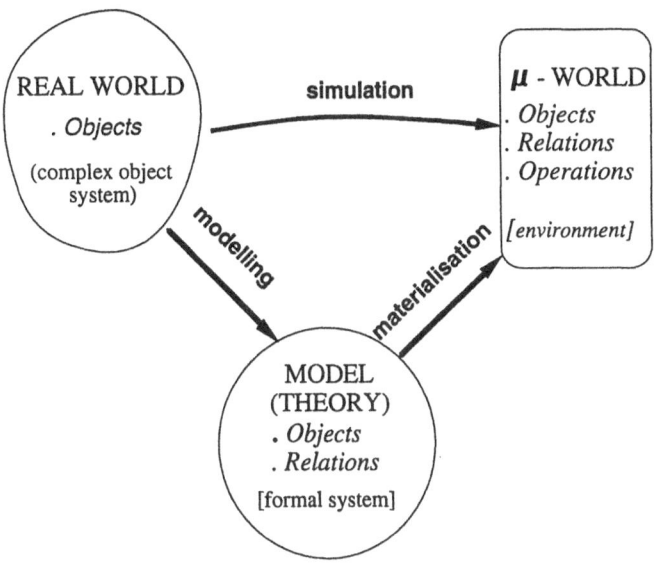

Figure 2

The mapping which allows one to pass directly from the initial system to the microworld is the only truly visible part of the user of the microworld, which then appears to be a simulation. The underlying theory at first remains absolutely transparent.

It should be noted that we give the term simulation here a slightly different meaning to that which tends, in everyday language, to be confused precisely with the notion of microworld. We would like to make a distinction here between microworld and simulator. Indeed it may be noted that a microworld charac- terised as above differs from a "simple" simulator in so far as the microworld allows more the creation of new objects presenting new properties. Moreover, as we point out that the use of a microworld or a simulator is always placed within the framework of learning (even if that is possibly only explicit to a greater or lesser extent[4]) it must be noted that a difference between "simple" simulator and microworld is established. Indeed, the contents of a learning (of a scientific nature are defined according to a certain theory: geometry with certain algorithmic aspects for a geometry microworld, ecology for a simulator such as SimEarth®, dynamics (centrifugal force) for a race track game. In a microworld the theory corresponding to the stakes of the learning is directly and largely involved in how the (computer) creation functions whereas this is not so much the case for a simulator.

In the preceding schema no particular place is given over to mathematical microworlds, which is why the definition proposed in paragraph 1 does not seem

---

[4] This is particularly sensitive in the framework of certain "computer games" connected to our subject.

to be restrictive, as [Lawler 89] on the contrary implies. In the case of mathematics however, one particular situation appears:

Although in general the two upper bubbles (the System of objects and the Microworld) stem clearly from materialism, as opposed to that of theory – hence establishing a parallel in their perception, a correspondence between two realities – this can no longer be the case in mathematics.

The first example is that of the microworld of calculation on numbers represented by a simple calculator with four operations, $\sqrt{\ }$ and %. Numbers are available, those which can be entered by the keyboard and six operators which allow others to be created. Relatively complex phenomena can be highlighted such as, for example, the equivalence of $\sqrt{1+x}$ to $1+\frac{x}{2}$ in the proximity of zero. The system of complex associated objects is that of numbers which, precisely, do not constitute a tangible reality for everyone.

Another example could be that of Euclidean geometry (i.e., which functions for itself and not as a modelling of the user's dimensional universe), here the arrow of mapping is reduced to identity and the microworld carries out the materialisation of the abstract (non significant) entities of theoretical geometry. This is a materialisation, in fact, the creation of a theory, and it is in this situation that mathematicians speak of model (in the sense of the theory of models).

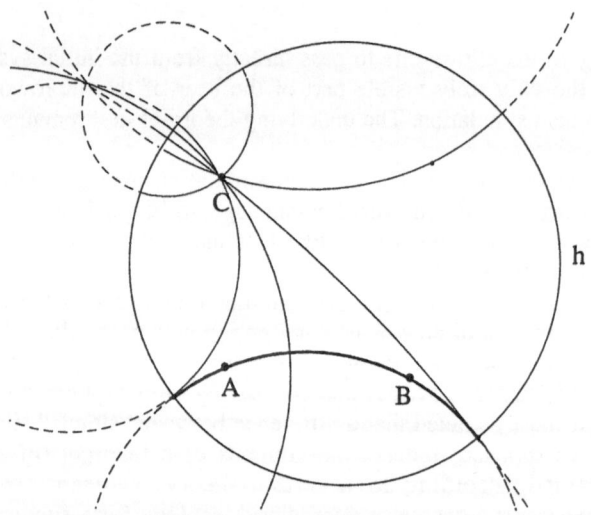

The points of the model are those inside a circle of the ordinary plane h; the "straight lines" are the part inside the circle h (infinite line), of the circles of the plane orthogonal to it.
We have drawn the "straight line" joining the two points A and B, dropped from C the "perpendicular" to (AB), and raised in C the "perpendicular" to the latter. We have also traced any parallel led from C to (AB) as well as the two limiting parallels.

Figure 3. Hyperbolic geometry (Poincaré's model)

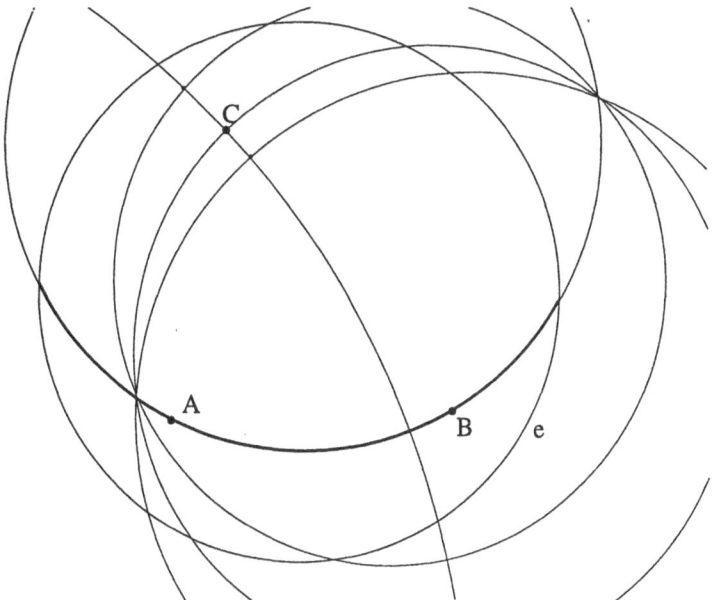

We consider a sphere, its poles, its equator and its great circles which are projected stereographically on the plane tangent to the south pole from the north pole: *e* is the image of the equator.

The points on the model are those inside the circle *e* (which plays the role of infinite line); the lines are the inside part of the circle *e* of the projections of the great circles.

We have drawn the "straight line" joining the two points A and B, dropped from $C \notin (AB)$ a perpendicular to (AB), and raised from C the perpendicular to the latter. We have also drawn two other "perpendiculars" passing through other points on the first, which all cut through (AB) at the same point.

Figure 4. Elliptic geometry (great circles model)

A given reality can be subject to several modellings, for example the trajectories of light can be the geodesics of a flat space (lines of Euclidean geometry with the postulate of parallels) or a curve. By limiting itself to the second dimension, geometry without the postulate of parallels can itself be subject to several materialisations in the framework of the analytical geometry of $\mathbb{R}^2$: the lines become the projections of the large circles of a sphere projected stereographically (elliptical geometry) or else the circles orthogonal to a circle of an ordinary plane (hyperbolic geometry). A microworld which is rich enough itself allows other, more complex[5] microworlds to be constructed.

---

[5] Or more generally: Cabri-géomètre can also be used for example in the creation of a simulation of LOGO by:

Numerous microworlds materialise through recourse to computers. Indeed, computers allow the behaviour of the initial system to be simulated through the implementation of the laws which govern it, as they are described in the theory. It can be seen that the functioning of a microworld depends on the theoretical model which alone allowed it to be created. The microworld functions on the basis of a theory which is supposed to be that (one of those) of the initial complex system. The pedagogical project which underlies the use of a microworld in a learning framework in fact consists in putting the user in the situation of appropriating for himself the contents of the corresponding theory. It must be emphasised once again that even if the ends of this theory occasionally turn up in the microworld, this does not appear explicitly to the user; the latter is confronted with a simulation of a complex system which he undertakes to study.[6] The analysis of the theory which underlies the functioning of a microworld and the conditions of its implementation (materialisation) is therefore a determining factor if the cognitive processes brought into play during a microworld session are to be accounted for. However, the microworld contains at its interface a part of the knowledge represented by the theory, represented by the choice of the operators. Once again we have the duality of syntax (system of signs) and semantics (realization of the theory).

We also encounter the case in which the implementation of microworlds rests on simply putting to work natural phenomena, the materialisation of which is supposed to be carried out by the microworld. This is the case with metal discs which move about on an air cushion.

The above examples emphasise that the concept of microworld is not necessarily linked to the concept of computer. In what follows we shall, however, look more closely at microworlds materialised by computers.

The Role of Direct Manipulation

The literature [Hutchins & al. 85] gives the work of [Sutherland 63] as one of the first creations based on direct manipulation; it is remarkable that this is a "sketchpad", that is the same concept as was at the origins of the Cabri project and which was also used in the Geometer's Sketchpad by [Jackiw 88]. In fact it

---

- the elimination of the elements of the menu belonging to Euclidean geometry, and
- the addition of "macros" representing the basic commands of any LOGO, forward, right-turn and left-turn.

In this way we obtain a direct-manipulation implementation of LOGO.

Moreover, in the environment of Cabri-géomètre, one can conceive real direct manipulation microworlds in the areas of optical geometry, algebra, and mechanics, ... It should be noted that this capacity offered by geometry – and which, fundamentally, is specific to it – of allowing such materialisations to be carried out would seem to have been only poorly used by the designers of CAD type systems.

[6] That every form of theory cannot be absent from a materialisation reflects the point stressed by the designers of Steamer which led them to favour simulations which guarantee a form of "conceptual fidelity" [Hollan & al. 84].

was only towards the end of the 1970s with the Star project at the Rank Xerox laboratories in Palo Alto that the original tentatives emerged from their confidential state. The essential idea of the Xerox Parc researchers who worked on the design of Man-Machine interfaces is based on the notion of metaphor, in this case the metaphor of the office, with its files, filing cabinets and tools. The electronic office of the Macintosh (known as multifinder) in fact appears to be a real microworld in which the objects are the filing cabinets linked up through files. The operators allow the hierarchy to be modified, and allow some to be destroyed and others to be created.

Direct manipulation is often linked to the principle of metaphor, and [Schneiderman 82 & 83] introduces the term to characterise interfaces with:
"• Continuous representation of the object of interest.
• Physical actions (movements and selection by mouse, joystick, touch screen, etc.) or labeled button presses instead of complex syntax.
• Rapid, incremental[7], reversible operations whose impact on the object of interest is immediately visible."

[Hutchins & al. 85] and more recently [Nanard 90] do not remain on this level which characterises a type of interface but introduce the user as a cognitive subject and as an element of appreciation[8]. The concept introduced is that of *direct engagement*. According to [Nanard 90] "The impression of direct engagement matches what the user feels when he can act directly and freely on the representations of the objects in his own world and perceive immediately their reactions."

LOGO, the archetype of the microworld, does not implement the concept of direct manipulation but this is probably due to the historical conditions of the development of the LOGO project. It was only at the end of the 1980s that most of the system environments for work stations started to use metaphors systematically, in particular the metaphor of the electronic office (Presentation Manager, OS/2, X-Window, Open Look, OSF-Motif,...)

Let us recall, because they are important, the principles which the designers of Star adopted [Smith & al. 82]:
"• familiar user's conceptual model,
• seeing and pointing versus remembering and typing,
• what you see is what you get,

---

[7] *Rapid* means that the time the system takes to react to the data fed in by the user must be short, *reversible* exists for the possibility of cancelling any order (to allow the user to undertake with confidence certain actions which can always be corrected if they do not suit), *incremental* means that each modification brought about by an individual feed-in remains small in relation to the whole of the task in progress.

[8] This is important because in this way any tentatives which are only redecorations of interfaces conceived independently of direct manipulation are disqualified, by adding an interaction based on graphics and mouse, without however satisfying the essential part of direct manipulation.

Windows 95, for instance, finally implements the trash metaphor to allow for deleting objects – this is 16 years after the Star machine and 12–13 years after the Lisa/ Macintosh from Apple. *Ideas move slowly even in computer sciences.*

• universal commands,
• consistency,
• modeless interaction,
• user's tailorability".

As well as "near-necessary" conditions in order for the impression of direct involvement to be produced, such as *semantic correspondence, articulatory correspondence, articulatory suggestivity, operational facility, reversibility,* [Nanard 90] also introduces, after [Hutchins & al. 85], the criterion of *free explorability.* This last point is important because it is linked directly to the possibility of the user, in a situation of learning through (possibly guided) discovery, to appropriate for himself the concepts which are new to him but which appear to be relevant to the solution of a problem, if, indeed, the interface makes them directly available to him. [Nanard 90] gives the example of "One of the great merits of MS-Word, [which] is indeed, thanks to its interface which is fairly close to direct manipulation – despite numerous awkward points – that it plays a pedagogical role in document manipulation". The consequences of this point of view certainly merit further study and analysis, especially in our perspective of learning environments based on microworlds.

Direct manipulation is a powerful concept [JM Laborde 95]; its use in an environment such as that of Cabri-géomètre radically transforms:
• the status of the figure which becomes a dynamic object,
• the status of the theorems which become the invariables of a figure (up to then they were more what the teacher asked the pupils to obtain through the subtle game of a series of reasons based on frequently inaccessible global architecture).

The initiators of the concept of direct manipulation were aware of its misleading aspect, which they summed up as follows: "A subtle thing happens when everything is visible: *the display becomes reality.* The user model becomes identical with what is on the screen." [ibid. p. 260]. This transfer force which may induce certain (possibly false) conceptions in the user has also been noticed among the users of Cabri-géomètre who are led to conceive a "Meccano" geometry in which, for example the points become articulations, certain straight lines become guides, in which geometry becomes mechanics.

3. A Few Problems in Geometry –
Conditioning of a Microworld by a Theory

It is interesting to note, in the case of different microworlds, how the schema indicated above works. For LOGO programming language, one could choose, for example, as a *system of objects* , that of the situation of calculations, and as a *theory* the algorithmic. The strength of LOGO stems, on the one hand, from the fact that it appears as a materialisation of the algorithmic in general whose strength it reflects. On the other hand, this situation does not allow LOGO to possess, as such, a wealth of knowledge. The situation, from the point of view of knowledge, becomes much more interesting when the system of objects is reduced to geometric movements, which is of course the current practice of the LOGO turtle.

A Few Characteristics of Geometry

Geometry has attracted numerous creations. It may be noted that this field is developing between two poles; on one hand we have a modelisation of our spatial environment, frequently reduced to two dimensions (geometry containing only one line with at least two points would not hold our attention for long). This modelisation creates a theory in which sensorial perception of the phenomena of everyday life plays an important role. On the other hand the theory obtained proves rich enough to raise problems which are essentially within the theory, which allowed Pascal to hold in great esteem "the behaviour of geometrical demonstrations, that is to say methodical and perfect"[9] This duality allows one to distinguish between two aspects of the geometrical figure, which have already been pointed out as being problematical from a didactic point of view [Parysz 88]: on one hand the figure as a drawing (Geometry of observation) and on the other hand as a drawing of arbitrary precision (Geometry of deduction). The microworld of geometry here provides a third aspect, with the concept of class of figure (or configuration).

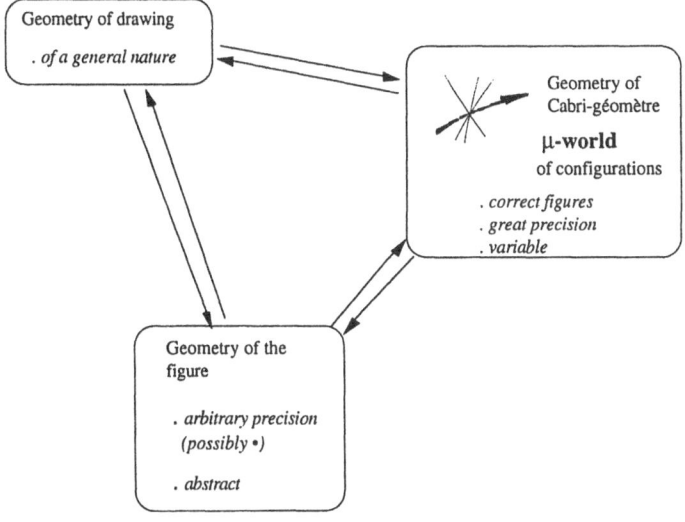

Figure 5

---

[9] Quote from the first of two short treatises entitled *De l'esprit géométrique et de l'art de persuader* (Of the geometrical mind and the art of persuasion), dating probably from 1657–58. In fact Pascal notes that the word "geometry" belongs to the genus and species in accordance with his time in which frequently the term of geometer was used for mathematician.

124   J.-M. Laborde

In Figure 5, the triangular layout recalls that of Figure 2. However, the geometrical figure appears in the microworld with some new characteristics, such as the fact that here the figures are variable. The state of the screen at each instant only provides one of the representatives of the figure, yet all the representatives (a configuration) are in the machine (see also [Bellemain 89]).

The drawing-figure duality has been analysed as being responsible for three snags (cf. [Yerushalmi & Chazan 90]).

- the particular nature of a diagram leads the pupil to take into account the characteristics of the diagram, which have no relationship with the properties of the figure and therefore are probably not adapted to the problem;
- the standard nature of certain diagrams brings out in the pupil the appearance of stereotypes, which later hinder the recognition of property in non-standard situations;
- "inability to 'see' a diagram in different ways", or in particular to grasp the whole and a part simultaneously.

On the last point we quote the example of the figure formed from a triangle *ABC* and of three points on its sides: *A'* on *BC*, *B'* on *CA*, *C'* on *AB*, with the hypotheses *B'C' // BC, C'A' // CA* and *A'B' // AB* as in Figure 6a.

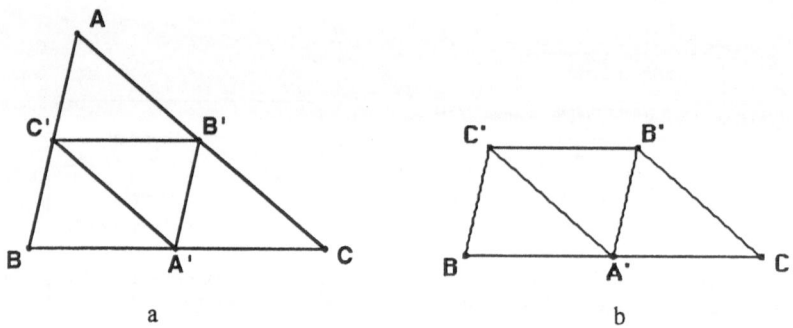

a   b

Figure 6

This example has been studied in [Duval 88] where it is reported that out of a population of pupils of about 16 years old, only 18% of them, managed to show that *A'* is necessarily the middle of *BC*, whereas 42% manage it in the case in which the figure is "shortened" as in Figure 6b.

The introduction of a geometry microworld such as Cabri-géomètre[10] provides a modification of the conditions for teaching geometry on these three points, with really new possibilities [Schumann 90], such as:

---

[10] Cabri-géomètre is certainly remarkable because of the determined way in which the principle of the dynamic modification of the figures is implemented, most other creations in the field also offering means to modify a figure by conserving its

- the figure loses its stereotyped nature;
- critical cases and border-line cases can easily be analysed,
- a single geometrical situation can be envisaged from different angles.

One should not, however, lose sight of the fact that the materialisation carried out is often far from being a model of the theory in the mathematical sense, that is that the theory that the user of a microworld could appropriate may be quite far from the one he had decided on. (Remember the somewhat caricatural example of the geometry of tore[11] induced by LOGO and let us consider that of Cabri-géomètre as well for example which can induce certain mechanistic-type conceptions in which the points are perceived as articulations, the straight lines are the guides of the "points on an object" see below).

Along the same lines it would seem obvious that any creation of Euclidean geometry cannot but present certain metric characters and that questions (of factor) of scale slip in unnoticed even though a priori the purely Euclidean "model" is exempt from them. There again the consequences of these phenomena on the didactic level are far from trivial.

A microworld does not solve as such these difficulties, it merely proposes a framework in which certain aspects of the difficulties encountered in the teaching can be managed differently by the teacher. Of course(?), the fundamental problems remain and the pedagogical use of microworlds is open to modellings following various cognitive hypotheses.

Finally, the duality mentioned is only an instanciation of that encountered in general in the couple physical reality-model to which microworlds appear to be systematically confronted.

4. What Type of Intelligence in a Geometry Microworld?

The Level of Interface

The illustrations below show some of the difficulties encountered for the representation on a bit-mapped screen of geometrical objects.(It may be recalled that the first LOGO creations did not use "pixel" screens but instead directly piloted an electron gun analogically.)

---

specification, which surely makes this point a didactic invariable. This includes such software as:
  - Euclide [Allard & Pascal 86], a language for plane geometry designed as an extension of LOGO, whose command language it has inherited,
  - Geometer's sketchpad [Jackiw 88], an American creation on Macintosh, which presents numerous similarities with Cabri-géomètre (direct manipulation),
  - Geometric Supposer [Schwartz & Yerushalmi 85], with command language
  - Geocon [Gerd Holland 89], a German product, Felix [Kadunz & Kautschitsch 90], an Austrian prototype, now (1995) released under the name Thales, Euklid (1994), a plagiary of Cabri-géomètre by Mr. Mechling (Offenburg, Germany), ...

[11] See note 2.

Figure 7 gives the graphic of a segment joining two points; it appears clearly asymmetrical, contrary to the user's spontaneous conception. Certain refinements giving a symmetrical look could be conceived.

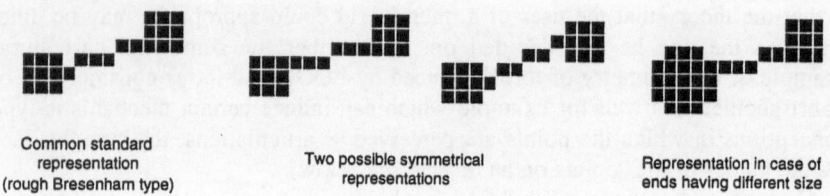

Common standard
representation
(rough Bresenham type)

Two possible symmetrical
representations

Representation in case of
ends having different size

Figure 7. Different screen representations of a short line segment (highly enlarged)

In the same way, it appears that if *A* and *B* are two points on a line defined elsewhere, the segment (the line) joining them does not correspond to the pixels of the original line, which again shocks the concept of line.

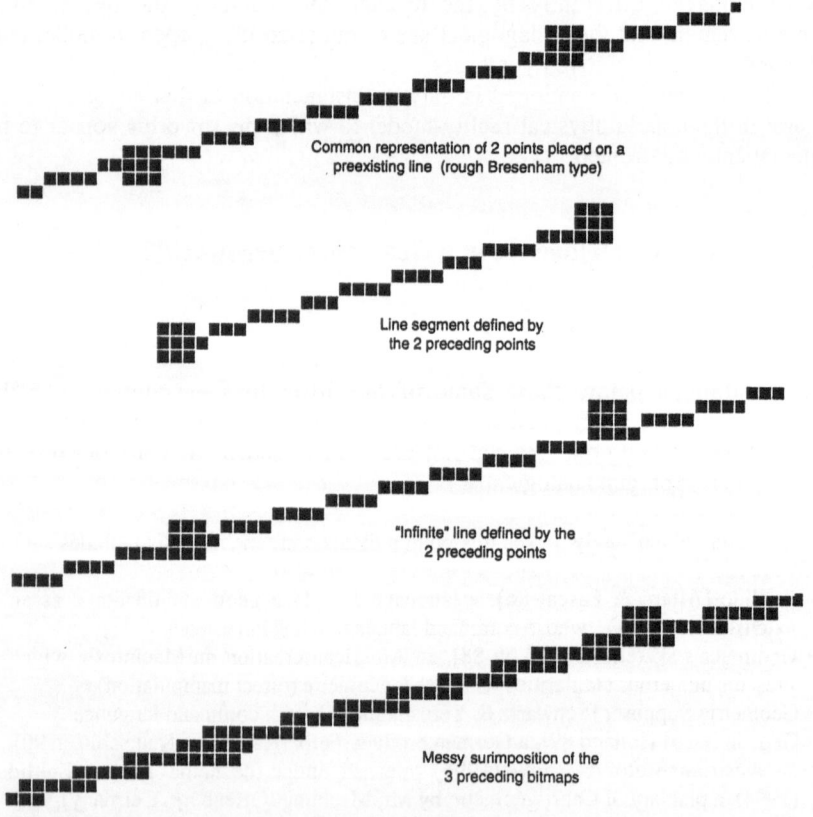

Common representation of 2 points placed on a
preexisting line (rough Bresenham type)

Line segment defined by
the 2 preceding points

"Infinite" line defined by the
2 preceding points

Messy surimposition of the
3 preceding bitmaps

Figure 8

In these exceptional cases a display with anti-aliasing could perhaps provide the necessary improvements although a more acceptable solution would consist in using knowledge other than the software's minimal knowledge of the objects it is manipulating. The display functions of CABRI should "know" that the representation of a line contains that of the segments whose ends belong to it.

In the same way and to a lesser extent on the interface, the software should "know", and not simply for graphic reasons, that the perpendicular led by $P \in D$ to D' which is itself perpendicular (by previous construction) to D, is of course D.

Here is a list of the situations in which knowledge of a geometrical nature would avoid duplications of objects during the construction of figures:

- perpendicular led by $P \in D$ to D' itself perpendicular (by previous construction) to D;
- parallel led by $P \in D$, to D;
- line or segment passing through A and B in which $A \in D$ and $B \in D$;
- creation of the middle of a segment on which one already has the intersection with its perpendicular bisector;
- perpendicular to Δ in A itself perpendicular to D even though we have already the parallel in A to D.

Management of borderline cases allows the bisecting line of two converging lines one of which becomes parallel to the other to continue to exist as the equidistant parallel.

More generally, the intelligent management of the intersection of two lines which are made parallel would allow the circle circumscribed to a triangle which has become flat, to continue to exist as the line containing the triangle or equally, would allow the mid-section of a segment whose ends belong to a circle to continue to be a diametre even if the ends of the segment meet.

We shall call this type of knowledge **on-board knowledge** as it conditions a function of the interface which can respond in any circumstances to the expectations of the user.

A more in-depth analysis of the on-board knowledge necessary to Cabri-géomètre is in progress.

In Cabri-géomètre, the "geometrical" point at the intersection of two lines for example, is not considered by the software, until it has been explicitly constructed by the user. This way of functioning, pedagogically justifiable (it is a "good" thing that nothing exists without having been defined) proved extremely problematic in particular during the experiments with the prototype HyperCabri (see below). Indeed, the interface breaks the impression of direct engagement as the user has the impression of finding himself in an environment in which he cannot do at least as much as he can do in a paper-pencil environment: this is due to a shortcoming of the principle of direct semantic correspondance. It would be useful to obtain a more subtle management of the interface, which would take into account the user as a cognitive subject, that is to create yet again a more intelligent interface.

Piloting a Microworld, the Modelling of the Learner

The review of the numerous problems raised by the use of a microworld as an element in an educational programme should not hide the fact that the idea of linking to a microworld an explicitly tutorial component, remains one of the most promising in the future of CAL. It should be remembered that recourse to a microworld may simply shift the problems and that for the moment research has not yet gone beyond the initialisation phase; we are far from declarations or unfounded optimism which would recall Anderson on The LISP Tutor[12] [Anderson & Reiser 85] "to have on campus by 1986 a personal computer with 1 megabyte of memory capable of 1 million instructions per second. ... These technological trends encourage optimism about the future of intelligent tutoring efforts, of which the LISP tutor is one. We hope that, with continued research in domains such as high school mathematics and college-level programming, we will soon establish the conceptual foundations to use the computational power that will be available: The prospect is great of providing every student with the educational benefits of a private human tutor. When this happens, the consequences for American education will be nothing short of revolutionary."[13] A great deal of hope has been invested by the protagonists of the concept of microworld as the solution to educational problems. However, in keeping with the acknowledgement that on occasion one witnesses a simple shifting of the problems, one may think that the simple virtue of encouraging exploration alone is not enough to modify the system of the user's knowledge. In fact we agree to recognise, on the basis of experience of Piaget, the central role given over to the problem in a learning process: "the solution to a problem is the source and criterion of knowledge" [Vergnaud 81]. In any use of LOGO the choice of problems is that of the teacher and only "good" teachers achieve "good" results. What would seem to obtainable to us is the intelligent running of a working session within a microworld environment.

With this in mind we have carried out a prototype experiment in which the Cabri-géomètre microworld was coupled with a tutorial model in the context of a task which had been the object of a preliminary didactic study: the construction by 12–14 year-old-pupils of a square from the data of one of its sides.

Most authors recognise that an intelligent tutor must include in particular three components, one for the representation of the knowledge of the field, another for the piloting of the didactic interaction and finally the most problematic component, the pupil model. In our prototype [JM Laborde & Sträßer 90], which functions on a particular task, the pupil model consists of an evaluation of his production (the figure he constructs is supposed to become a square), as concerns its explicit and implicit geometrical properties. According to this evaluation the

---

[12] This paper's promisingly modest subtitle is "It approaches the effectiveness of a human tutor".

[13] Ten years later, the basic power of PCs has been augmented by a factor of 30 million [JM Laborde 95]. Such an augmentation is still not visible in the efficiency of most ITSs. This means that research is still far behind and more is needed.

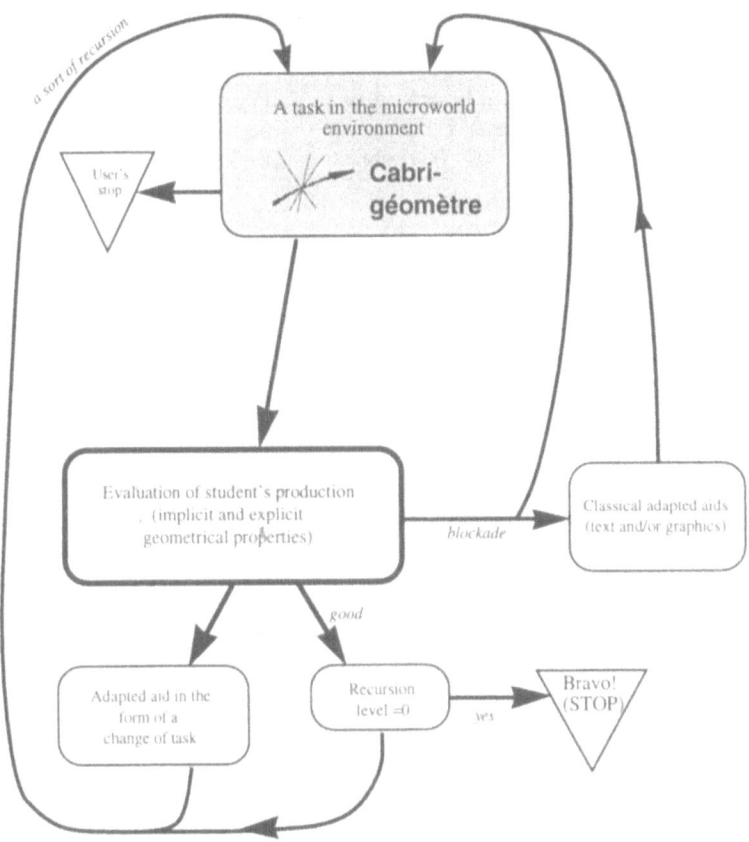

Figure 9. Overall architecture of HyperCabri

system proposes to the learner other tasks to be solved in a more subtle process than the ordinary decomposition of a problem into sub-problems. As an example, the system would propose to the learner who has failed on the question of the construction on the perpendicular of the initial side, of a side of the same length, the equivalent task but in the case of a more acute angle than a right angle (this is a variation of the variable of the corresponding task). Figure 9 provides the general architecture of the system.

The didactician's work has provided a task, presenting a certain number of the desirable characteristics for an activity which would ensure the cognitive progress of the subject:

- The task is simple but open to numerous strategies (a preliminary study listed at least eight which were effectively used);
- the retroactions (visual feedback) of Cabri-géomètre are not ambiguous, the shifting of one of the ends of an initial segment will not leave the figure with a square shape unless the construction is correct;

- some of the stages of the construction may, in case of failure, be presented in a different or more standard form allowing the user to go beyond his initial skills (learning);
- the user's procedures can be traced even if they only act on the implicit or simply intentional elements.

See [Capponi 90, C Laborde 90, JM Laborde & Sträßer 90] for an in-depth study of these different aspects.

Conclusion?

It would no doubt be interesting, on the basis of the most important points underlined in this contribution, to indicate what the "essential specifications" of a calculus microworld dedicated to arithmetic and algebra could be, a future "Cabri-algèbre". Indeed, if the concepts presented above are robust enough, it should be relatively possible to sketch such informal specifications:

- The system of objects is that of calculus.
- The theory is that of arithmetic, algebra, some analysis. In the same way as Cabri-géomètre is a world for the construction and manipulation of geometrical figures and for the appropriation of the properties of geometry, Cabri-calculus would be a world for the construction , manipulation and evaluation of algebraic expressions, as well as for the appropriation of the properties of calculus.
- the operators would have to include
  - possibilities for the transport and substitution of expressions,
  - tools for the carrying out breakdowns (factors, prime factors, simple elements, limited developments,...),
  - the implementation of set operations,
  - the composition of operators (macros, e.g., for LCM and GCD),
  - a direct manipulation grapher (action on axes, scales,...).

The fundamental metaphor would still be that of the sketchpad (CABRI) and the interface could include aspects of hypertext.

One of the characteristics of Cabri-géomètre is the visual feedback matching each modification of the figure being elaborated. What could be the counterpart for this algebra microworld? Perhaps one could imagine this through the graph drawer as the quantitative guarantee of the qualitative adequation of the operations being undertaken?

We hope to have marked in this document the careful optimism of the future prospects for learning environments: we are at the dawn (still at the dawn) of surely important progress, the revolution in education and computers will indeed take place, even if the research to be carried out is considerable (in didactics, in computers, in AI, in cognitive sciences, ...), while profitable uses are immediately on the agenda especially if one manages to truly insert the computer as a supplementary pole in the classical schema of the Student–Teacher–Knowledge interaction.

References

Allard J.-C., Pascal C. (1986). Euclide, un langage pour la géométrie plane, logiciel et manuel, Cedic-Nathan Paris.

Anderson J. R., Reiser B. J. (1986). The Lisp Tutor, Byte 10(4), 159-175.

Baulac Y. (1990). Un micromonde de géométrie, Cabri-géomètre, thèse de l'Université Joseph Fourier Grenoble.

Baulac Y., Bellemain F., Laborde J.-M. (1988). Cabri-géomètre, un logiciel d'aide à l'enseignement de la géométrie, logiciel et manuel d'utilisation, Cedic-Nathan Paris.

Bellemain F. (1988-89). Le logiciel "Cabri-géomètre", un nouvel environnement pour l'enseignement de la géométrie, Publications de l'Institut de Recherche de Mathématiques de Rennes, (5) Didactique des Mathématiques

Bellemain F. (1992). Conception, réalisation et expérimentation d'un logiciel d'aide à l'enseignement de la géométrie, Cabri-géomètre, Thèse. Grenoble, Université Joseph Fourier.

Boder A. (1990). Ramos, Exposé aux journées Méthodologie de production de moyens d'enseignement, Dep. de l'Instruction Publique du Canton de Genève (centre EAO) Grimmialp Suisse.

Brown J.S., Burton R.R. (1979). An Investigation of Computer Coaching for Informal learning activities, International journal of man-machine studies, 11, 5-24.

Capponi B. (1990). Probleme tutorieller Geometrie-Systeme am Beispiel des Prototyps HyperCabri, Report, Institut für Didactik der Mathematik (IDM) University of Bielefeld, Germany.

Duval R. (1988). Pour une approche cognitive des problèmes de géométrie en termes de congruence, Annales de didactique et de sciences cognitives, IREM et Université Louis Pasteur Strasbourg (1), 57-74.

Forbus K. (1984). An Interactive Laboratory for Teaching Control System Concepts, BBN Report 5511, Bolt Beranek and Newman Inc., Cambridge, MA.

Hollan J.D., Hutchins E.L., Weitzman Louis (1984). STEAMER: an interactive inspectable simulation-based training system, AI Magazine 5(2), 15-27.

Hutchins E.L., Hollan J.D., Norman D.A. (1985). Direct manipulation interfaces, Human-Computer Interaction (1), 311-338.

Holland G. (1989). Geocon, Karl Liebig University, Giessen, Germany.

Jackiw N. (1988). Geometer's Sketchpad: The Visual Geometry Project, Swarthmore College, USA, and KeyCurriculum Press, Berkeley, CA (1990-1995)

Kadunz, Kautschitsch Herman (1990). Felix, Kärntner Symposium für Mathematik-didaktik, 1990 Klagenfurt, and Universität für Bildungswissenschaften Klagenfurt, Austria.

Laborde J.-M. (1985). Projet d'un Cahier de Brouillon Informatique de Géométrie, Rapport interne LSD (IMAG), reproduced in [JM Laborde 95].

Laborde J.-M., Trilling L. (1989). Conception et réalisation d'un système intelligent d'apprentissage de la géométrie, rapport de Projet IMAG.

Laborde J.-M. (1989). Designing Intelligent Tutorial Systems: the case of geometry and Cabri-géomètre, IFIP WG 3.1 Working Conference on Educational Software at the Secondary Education Level, Reykjavik.

Laborde J.-M., Sträßer R. (1990). Cabri-géomètre: a Microworld of geometry for guided discovery learning, Zentralblatt für Didaktik der Mathematik 5, 171-177.

Laborde J.-M. (1995). Des connaissances abstraites aux réalités artificielles, le concept de micromonde Cabri, Environnements Interactifs d'Apprentissage avec Ordinateurs (tome 2) Eyrolles Paris 5, pp. 29-41.

Laborde C. (1990). Zur einer Didaktik des Geometrie-Lernens unter Nutzung des Computers, Rapport de l'Institut de Didactique des Mathématiques (IDM) University of Bielefeld, Germany.

Lawler R.W. (1987). Learning environments: now, then, and someday. In: R.W. Lawler & M. Yazdani (eds.), Artificial intelligence and education, pp. 1-25. Ablex, NJ.

Minsky M., Papert S. (1969). Perceptrons: An Introduction to Computational Geometry, MIT Press, Cambridge, MA.

Minsky M., Papert S. (1972). The '72 Progress Report. AI Lab. MIT, Republished under the title Artificial Intelligence by University of Oregon Press.

Nanard J. (1990). La manipulation directe en interface homme-machine, Thèse de Doctorat d'Etat, USTL Montpellier.

Papert S. (1980). Mindstorms: Children, Computers, and Powerful Ideas. Basic Books, New York.

Parzysz B. (1988). Problems of the Plane representation of Space Geometry Figures, Educational Studies in Mathematics, 19(1), 79-92.

Pascal B. (1657). De l'esprit géométrique et de l'art de persuader, in Œuvres complètes, Ed. du Seuil Paris (1963).

Schneiderman B. (1982). The future of interactive system and the emergence of direct manipulation, Behavior and Information Technology (1), pp 237-256.

Schneiderman B. (1983). Direct manipulation: A step beyond programming languages, IEEE Computer (16)8, 57-69.

Schumann H. (1990). Neue Möglichkeiten des Geometrielernens durch interaktives Konstruieren in der Planimetrie: Der mathematische und Naturwissenschaftliche Unterricht (4), 230-242.

Schwartz J., Yerushalmi Michal (1985). The Geometric Supposer, Sunburst Communications, Boston, MA.

Smith D.C., Irby C., Kimball R., Verplank W., Harslem E. (1982). Designing the Star User Interface, Byte 7(4), 242-282.

Stevens A.L., Roberts B., Stead L. (1983). The Use of a Sophisticated Interface in Computer-Assisted Instruction, IEE Computer Graphics and Applications, 3, 25-31.

Sutherland I.E.: Sketchpad (1963). A man-machine graphical communication system, Procedings of the Spring Joint Computer Conference, pp. 329-346.

Thompson P.W. (1987). Mathematical Microworlds and Intelligent Computer Assisted Instruction. In: G. Karsley (ed.) Artificial Intelligence and Instruction, Addison Wesley, Reading MA.

Vergnaud G. (1981). Quelques orientations théorique et méthodologiques des recherches françaises en didactique des mathématiques, Recherches en didactique des mathématiques, (22), 215-231.

Yerushalmi M., Chazan D. (1990). Overcoming Visual Obstacles with the Aid of the Supposer, Educational Studies in Mathematics, 21(3), 199-219.

White B.Y., Frederiksen J.R. (1985) . QUEST: Qualitatitive understanding of electrical system troubleshooting, ACM SIGART Newsletter 93, 34-37.

Winograd T. (1972). Understanding Natural Language, Academic Press, New York.

A Constructivist Model for Redesigning
AI Tutors in Mathematics

Richard Lesh[1] and Anthony E. Kelly[2]

[1] Professor of Mathematics, University of Massachusetts at Dartmouth
North Dartmouth, MA 02747, USA, E-mail: rlesh@nsf.gov

[2] Associate Professor of Education, Rutgers University
New Brunswick, NJ 08903, USA

Introduction

This chapter suggests a new approach to designing AI tutors that grew primarily out of results from a study that is reported in greater detail in a paper entitled *The development of tutoring capabilities in middle school mathematics teachers* (Lesh & Kelly, in press). That study focused on a ten-week project in which real human tutors provided the intelligence behind a computer-based tutoring system. Our goal was to investigate the understandings, assumptions, and procedures that were really used in tutoring by teachers who were at various levels of expertise – and to observe the evolution of these tutoring abilities over time, as tutoring effectiveness gradually improved.

A brief description of the preceding study is given in Appendix A. For the purposes of the present chapter, the main result that we want to stress is that, during the tutoring sessions when the tutoring was least effective, the teachers generally operated according to rules of the type that have characterized traditional types of AI-based tutors. On the other hand, during sessions when tutoring was most effective, the preceding principles tended to be explicitly rejected – and were replaced by the kind of constructivist principles that will be described briefly in this chapter.

Overall, our studies suggest that limitations inherent in some of traditional AI tutors arise not only because of over-simplifications arising from attempts to preserve programming convenience and ease of computer modeling, but also because assumptions are made that mis-characterize the nature of mathematics and mathematics learning.

In this chapter we review the traditional model of AI tutor design, and question in turn its underlying assumptions. Then, we suggest in its place an approach to tutoring that places the need for intelligence not so much in the machine, but rather in the off-line preparation of structurally rich, open-ended environments that encourage students to construct specific models that have time-tested utility and conceptual power. Such environments evoke the intelligence of the students during run time, allowing them to evoke mathematics as a rigorous modeling language; rather than having them be constrained by the machine to follow its notion of the best (procedural) path. This constructivist view of mathematics learning clearly is not one that can be fully or easily modeled by a computer, but the apparent loss in terms of machine omnipotence is more than compensated for by the demand that students be responsible for and be able to defend their own constructions in mathematics. It is also a view that is more consistent with the kinds of conceptions of mathematics, mathematics learning, and mathematics problem solving that are being expressed in documents such as the National Council of Teachers of Mathematics' *Curriculum and Evaluation Standards for School Mathematics* (NCTM, 1989), or the Mathematical Sciences Education Board's *Reshaping School Mathematics* (MSEB, 1990), or the American Association for the Advancement of Sciences' *Project 2061* (AAAS, 1989, 1990).

AI Tutors and Human Tutors: A Comparison

Traditional AI tutor architecture involves: (i) a system of condition-action rules written to characterize expert and novice states of knowledge, and (ii) a set of rules written to generate leading questions and advice to guide students from givens to goals within well-defined problem spaces – while monitoring and evaluating each step to ensure that students never stray too far from the tutor's notion of the (only) correct solution path(s). Stated more explicitly, the steps involve:

1) determining the exact nature of the student's knowledge,
2) comparing the preceding novice-level knowledge with (a singular and static conception of) expert-level knowledge,
3) asking directed sequences of (tiny and precise) questions that represent a correct next step for moving continuously and incrementally from givens to goals,
4) parsing, interpreting, and evaluating the student's responses,
5) giving the student immediate feedback about whether or not responses are correct,
6) providing leading hints, helps, or follow-up questions if the student fails to learn.

Basic assumptions are made that: (i) relatively simple-minded *production system* models are adequate to produce simulations of human behaviors, and (ii) the knowledge that is required depends primarily on lists of rules. This conception of instruction has a number of advantages, particularly the ability to be precise and rigorous about the procedures in which the student is being instructed, and the student's relative position in the knowledge base. It also demands precision in the description of the process of tutoring. Traditional AI tutoring systems were seen as first steps in a promising direction. Technical difficulties (which were expected to disappear soon) were considered the main factors preventing effective execution.

Unfortunately, the preceding approach to building tutors has produced remarkably unimpressive results in terms of students' learning gains. For example, a significant number of subject matter experts in mathematics have expressed concerns that traditional expert-novice approaches to AI-based tutoring in mathematics (the focus of this chapter) tend to be based on questionable assumptions about the nature of mathematics, the nature of realistic problems, and the nature of desirable mathematics teaching (e.g., Kaput, 1990; Schwartz, 1990; Thompson, 1989). Indeed, our work suggests that the principles underlying traditional AI tutoring are among the least highly regarded by human tutors; whereas, more open-ended and constructivist approaches are the most highly regarded.

In effective human tutoring (Lesh & Kelly, in press) terrain's of knowledge were explored in such a way that the student was usually in the role of telling the computer (or tutor) what to do, and then evaluating computer-generated (or tutor-generated) results, rather than the other way around.

Furthermore, as our human tutors developed progressively greater expertise:
- Relatively little time was spent trying to diagnose or correct students' procedural bugs.
- Multiple linked representations were emphasized as powerful instructional devices; and representational fluency was encouraged on the part of students.
- Tutor's questions were often fuzzy in the sense that they encouraged multiple types and levels or interpretations and responses – and therefore tended to be self-adjusting in difficulty.
- Students were encourage to use computer-based procedure-executing tools (such as graphics-linked calculators and modeling tools) to leap-frog over unproductive procedural details.
- Within pre-planned and carefully structured exploration environments, tutors focused on following and facilitating students' thought processes, rather than the other way around.
- Students' errors were often simply ignored; or, in other cases, they were actually induced in order to confront students with the need for conceptual reorganizations.
- The objectives that were emphasized shifted away from basic facts, rules, and skills toward deeper, broader, and higher-order understandings and processes.
- The *idealized model of potential student knowledge* that our teachers used to direct their tutoring activities was not considered to be a description of the state of knowledge for a given student, nor was it a description of the state of knowledge of an expert in the given conceptual neighborhood. Instead, it represented a deep understanding of an elementary topic area.

AI Tutors: A Critique of Some Past Assumptions

We will now consider in turn a number of assumptions of AI-based mathematics tutoring systems, and state our disagreements with them.

Concerning the nature of instructional objectives: In general, to conform to the representations permitted within a rule-based production system, the objectives of instruction are assumed reducible to strings of relatively simple declarative knowledge (e.g., basic definitions, facts, and skills) plus perhaps a few higher-level managerial strategies (i.e., further rules) that focus on helping students decide: (i) when to apply a given fact, skill, or definition, and (ii) which fact, skill, or definition to choose in a given situation.

We believe that systems that emphasize procedures and declarative facts – the ones most amenable to rule-based systems – are ones that promote outdated assumptions of mathematics, i.e., that *i) mathematics is a fixed and unchanging body of facts and procedures, and (ii) to do mathematics is to calculate answers to set problems using a specific catalogue of rehearsed techniques* (MSEB, 1990, p. 4). We believe that an emphasis on skill and syntactic manipulation distracts the student from becoming involved in the larger task of using *mathematics as a modeling language*. In general, the philosophy of tutoring that will be described in this chapter is consistent with the Mathematical Sciences Education Board's view that:

The teaching of mathematics is shifting from an authoritarian model based on transmission of knowledge to a student-centered practice featuring stimulation of learning (MSEB, 1990, p. 5). *Teachers should be catalysts who help students learn to think for themselves. They should not act solely as trainers whose role is to show the right way to solve problems. The aim of education (is) to wean students from their teachers.* (MSEB, 1990, p. 40)

Concerning the nature of givens and goals for individual problems: To facilitate computer based answer-checking, answers to problems are assumed to have only a single type and/or level of correct answer. All of the information that is relevant to the solution of a problem is assumed to be readily available in a form that requires little generation, organization, or interpretation (e.g., coding) in order to fit the machine's available processing capabilities and knowledge representation system.

We believe that the above approach denudes the realistic task of mathematical problem solving. Successful problem solving depends also on factors such as the amount of information to take into account, the reasoning pattern on which answers are produced, and the recognition of conditions that might influence the validity or precision of the results. Therefore, the kinds of modeling processes that will be of interest in this chapter involve more than simply getting from givens to goals using clearly specified sets of procedures.

For problems and decisions in which the construction of an interpretive framework is critical, initial conceptions of givens and goals are often based on impoverished and distorted interpretations of the problem situations. Therefore, students (or AI tutors) who are preoccupied with the question *What should I do next?* often tend to do poorly compared with students (or human tutors) who pay more attention to the consideration of alternative ways to model (describe, explain, or interpret) the situation.

Concerning the nature of solution paths: Attempts are made to restrict the problem space (i.e., givens, goals, and permissible solution paths) so that only a small number of alternative solution paths are permitted. Therefore, tools such as calculators or graphic modeling software tend to be prohibited, and students are not encouraged to use multiple interacting notation systems (e.g., such as linked or interacting systems involving written symbols, spoken language, graphic diagrams, manipulatable objects, or experience-based prototypes). Also, the problems that are emphasized tend to be decontextualized so that students prior experiences does not produce a variety of ways to think about the situation (e.g., alternative givens, goals, and solution paths).

Restricting the problem space for the sake of handling the combinatoric explosion of options generated by many mathematics problems (e.g., consider the many paths leading to a solution of a complex equation, or the variety of combinations of numbers and operations that can produce a given number) hinders the students exploration of a problem. Restricting a student's use of tools solely because their use and contribution to the solution cannot easily be monitored by the machine simply produces constricted problem-solving environments at odds with the behavior of professional mathematicians.

Concerning the individualization of instruction: A naive view suggests that the primary way to individualize instruction is to design alternative versions of each instructional activity, with each alternative aiming at a student aptitude (e.g., which may be classified into categories such as: spatial, analytic, or other aptitude or learning style categories) .

The task of matching specialized lessons with specialized student needs (even those measured by reliable aptitude tests) may be an impossible (see Cronbach & Snow, 1977). Further, a multitude of factors, many of them unique to the individual at a given time (e.g., affect, retroactive inhibition, fatigue, motivation problems) will distort attempts at matching that are too fine grained. The general constructivist view of learning adopted in this chapter rejects simple-minded, static student modeling.

Concerning the monitoring of progress toward solutions: In order to ensure that students never stray too far from the tutor's notion of the (only) correct solution path(s), it is assumed to be necessary for the computer immediately to approve or reject each step that students take. This computer action is often a programming convenience because of the combinatorics of permitting students errors to evolve. But, immediate feedback is not available in much realistic problem solving. Further, it may minimize the chances that students will be able to monitor their own progress – and be able to judge for themselves the usefulness of intermediate results – thus promote passive problem solving.

Concerning the diagnosis of students' errors: Students errors tend to be attributed to faulty rules (e.g., procedural bugs) that are assumed to be repairable by deleting and inserting isolated components.

To be overly concerned with error models may be to lead students on paths along which errors are avoided. Or, in extreme cases, to present a combination of leading questions and instant approval of responses that lead students to a correct answers even though the students may not have understood the quantitative relationships in the problem – or why the various corrective steps were taken. This view of errors (again, due often a programming convenience) over-emphasizes procedural errors at the expense of having the student recognize that holistic reasoning patterns must be gradually constructed, differentiated, and refined (e.g., perhaps through students' structured interactions with their environments – which might include manipulatable concrete materials, interactive computer graphics, or interactions with peers or a combination of these factors). When one bug is fixed in a superficial way (i.e., without altering underlying problem interpretations or reasoning patterns) another bug tends to crop up to take its place – because the real source of difficulty had to do with the stability of a holistic conceptual system more than with missing or damaged system fragments (Bunderson, 1983; Sleeman, Kelly, Martinak, Ward & Moore, 1989).

Concerning the generation of helps: As students attempt to solve individual problems, help tends to be provided largely by generating the tutor's notion of the (only possible) correct next steps, thereby conveying the impression that alternative steps might be inappropriate. Computer advice to students that merely places them on the straight and narrow solution path ignores the reality of the evolution of mathematics in which many apparently plausible paths are explored that must later be abandoned (Polya, 1984). Giving students the impression that mathematics involves nothing other than mastering time-worn solutions promotes passivity on the part of the student.

Concerning the generation of follow-up problems (or assignments): Students assignments (including selections of problems within sequences or branching networks) are based on the tutor's notion of the (only possible) next step for the purpose of optimizing learning progress. Within a strict rule-based procedural approach to mathematics instruction, the tutoring system is limited in the type of tasks it can set. It will have no way of assessing that the student may fundamentally misunderstand aspects of the original problem that are conceptual and semantic. Further, the phenomenon that occurs in tutoring is a sort of psychological version of Heisenburg's famous *indeterminacy principle* in physics. Judgements of students capabilities always involve an interaction between: (i) the internal structures that the students use to interpret the situation, and (ii) the structure of the environment that the tutor creates. Therefore, the observation is always influenced by the observer and continually evolves. AI-based tutors have tended to neglect enlisting students as partners in making instructional choices and identifying instructional goals; and, they have tended to operate as though they knew how to teach better that the students knew how to learn.

Concerning the nature of expert vs. novice characterizations of knowledge: Inherent in the traditional AI tutor design is an implicit expert-novice continuum, along which the student can be moved, using rules, until the state of expert is realized by the student. The traditional novice-expert model posits static, polarized novice and expert end-points on a continuum of knowledge. The intermediate points on the continuum are seen as additive steps to greater sophistication. In subtractive terminology, therefore, the task of the tutor is to move students from their current Novice state (N) to an Expert state (E) by bridging the difference of $(E - N)$ by a matter of adding and deleting rules.

We believe that this subtractive model of tutoring is fundamentally flawed. The continuum it is proposed to traverse is epistemologically unsound because

(i) it is practically impossible to arrive at static descriptions of canonical novice and expert states when dealing with human subjects, and (ii) the development from novice to expert is not continuous. Novices exhibit models that are either unstable (e.g., Sleeman, Kelly, Martinak, Ward & Moore, 1989) in which case they do not lie on any identifiable continuum, or stable (e.g., in some areas of science, Clement, 1982; diSessa, 1982, 1983, 1987; McCloskey, 1983; Minstrell, 1982), in which case they clearly do not reside on the same continuum that describes the expert. On the other hand, experts do not always agree among themselves. They reflect upon their rules and models. They think *about* those models that they think with in response to demands of the environment, which in turn may re-characterize the environment for them. This process inevitably leads to disagreement in advancing areas of study.

Experts and novices not only do not share the same amount of knowledge and information they also understand it very differently, and the differences between these two poles are not additive. Therefore, in our building of tutors, we must forego a static expert model that merely describes expert knowledge. Instead, we must focus on a dynamic and evolving expert/environment coupling that is the essence of expertise.

We maintain that the flux that characterizes the attempts of experts to wrestle with the complexities at the leading edge of a field is similar to the flux that characterizes the attempts of novices along the entire continuum in their development of a fuller understanding of a field. Instead of a continuum along which growth is additive, we favor an evolutionary one along which progress is characterized by false starts, discontinuities, acceptance of positions that may be later reversed or abandoned, insight, creativity, paradigm shifts, and the awareness that the process of learning is fluid and without discernible end-point. In short, we believe that it is neither necessary nor desirable to define a single, remote, best state of knowledge in order to identify directions that will be better for a given student at any stage of development. Further, we believe it is preferable for students to tell computers what to do and to evaluate the computer-generated results, rather than for the computer to tell students what to do and to evaluate the student-generated results (Schwartz, 1989).

The central questions in designing tutors then become, (i) How it is that, at a given level of development, students are able to detect the need to develop a more refined way to think about the situation? (ii) How is it that students are able to develop beyond the limitations of their own initial conceptualizations of the problem situations? Clearly, students self-tutoring activities are not based on bridging gaps between expert and novice characterizations of knowledge.

From Tutoring as Procedure Execution
to Tutoring as Model Exploration

Just as it is undesirable to reduce the goals of K–12 mathematics to nothing more than low-level facts and skills (perhaps supplemented by a small number of higher-order managerial strategies), it is not desirable to try to reduce mathematics tutoring to lists of simple-minded rules and procedures. Mastering a flexible system of rules involves far more than one-at-a-time clarification of the meanings of each. It also involves learning how the various rules are related to one another, and it involves being able to use the rules within a flexible and adaptable system. It also involves learning about the conditions under which each rule should (or should not) be used or modified – as well as learning about possible ways that the rules can be modified and adapted to fit different kinds of circumstances. As in other problem solving and decision making domains, such lists of rules tend to be rather worthless if available interpretive models furnish only barren and distorted information about the situations in which the rules are intended to be used. We must ask instead, What is a good problem for tutoring?

Concerning the nature of realistic problems: According to the perspective that is consistent with our view of good tutoring, a good problem is one that requires model building. We refer to such problems as *realistic* problems. In such problems, an overwhelming amount of relevant information often is available, which needs to be filtered, weighted, simplified, organized, and interpreted before it is useful (Naisbitt & Aburdene, 1990). Or, meaningful patterns or relationships in the data need to be modeled so that rapid decisions can be based on a minimum set of cues. A model gives a holistic interpretation of the entire situation, including hypotheses about objects or events that are not obviously given (or that need to be actively sought out or generated).

Models allow the student: (i) to anticipate real events, (ii) to reconstruct past events, or (iii) to simulate inaccessible events. The task of problem solving has to do with the constructing, analyzing, critically assessing, and refining a model that conforms to Einstein's famous dictum that *every explanation should be as simple as possible, but no simpler!*

Realistic problems also tend to involve complex modeling cycles. For example:

• Several modeling cycles may be needed (with modifications and refinements to the model at each cycle) before the model is able to produce useful predictions in the modeled situation – i.e., before a sufficiently good fit is established between the model and the modeled situation.

• A series of models may be needed to solve a given problem. (i) The first model may involve a schematic diagram of the problem situation. (ii) The second model may involve a system of equations describing the most important relationships in the diagram. (iii) The third model may consist of a Cartesian graph of the preceding equations. (iv) Then, this third model may be used to make predictions about the original problem situation.

• Several (perhaps partly incompatible) models may need to be used in parallel. For example, in physics, to accurately describe the behavior of light, both a wave model and a particle model may be needed. Each model describes or clarifies only some aspects of the modeled situation – and ignores or distorts other aspects. Therefore, a combination of several models may be needed to produce useful results.

• Instruction is not so much a matter of constructing completely new models as it is a process of coordinating, differentiating, integrating, extending, refining, or modifying models that already exist. For most of the basic conceptual models that are reasonable to try to help students construct at a given level, most will already be at some intermediate stage of development. Or, if students have not already constructed some relevant concrete/intuitive/restricted form of a given model, then ideas that depend on this model are likely to be as inaccessible as Piaget's formal operational concepts tend to be for pre-concrete operational children. Yet, precisely because the most basic conceptual models in any field involve patterns that are useful in a broad range of situations, students usually enter instruction with more than blank slates of knowledge about them. Therefore, if the underlying reasoning pattern is accessible at some level of sophistication (perhaps in some restricted context).

Concerning the nature of mathematical models: The modeling view of mathematics that will be adopted in this chapter is based on the notions that (i) mathematics is not simply a static list of facts and rules (MSEB, 1990); (ii) mathematics is the study of a symbol system embedded in (usually several) external representational systems; and the meanings of internal models are influenced by unstated experiences, assumptions, and purposes of particular humans – as well as by the structural characteristics of the situations they have been used to interpret. These structures are (potentially) useful for making sense of real experiences (by describing, explaining, and predicting experiences); (iii) humans are not simply neutral observers of reality, and mathematics is not simply extracted from a reality that is completely external. Instead, fundamental models for making sense of experiences tend to be based on human constructions that are imposed on reality ; and (iv) as more sophisticated models are used to interpret experience, the nature of the *world of experience* is itself transformed in fundamental ways (Winograd & Flores, 1985). New meaningful patterns are continually being projected into reality, and the need for continuing rounds of conceptual development remain ever present.

According to the preceding points of view, the objects (equations, number systems, formulas, graphs) that mathematicians study all can be referred to as models for describing, explaining, or predicting real experiences.

Or, if we want to emphasize the underlying structural characteristics of these models, rather than emphasizing the concrete/perceptual characteristics of the objects in which they are embedded, they can be thought of as structural metaphors – or as patterns that humans embed within various types of notation systems (whose elements may be written symbols, spoken words, pictures, or concrete objects).

• Mathematical models are developed locally. Once a given model is constructed, it does not necessarily follow that other situations will be recognized in which the model could be applied. This is why the initial construction of a model generally represents only a small portion of the activities needed to endow it with generalizability and power. Even after a model has been constructed, it is usually necessary to: (i) analyze underlying assumptions and possible sources of errors, (ii) investigate characteristics such as robustness when various assumptions are violated, and (iii) modularize, unpack, and reassemble the model in a variety of ways so that it can be easily transported, modified, and extended to fit circumstances beyond those in which it was originally created. In other words, to increase the power and generalizability of a model, it is important to go beyond reasoning with the model, and also reason about it.

• Mathematical models are situated. To achieve maximum power and accuracy when mathematical models are used to describe, interpret, and explain realistic situations, the principles on which these models are built usually must capitalize on situation-specific factors, as well as on abstract reasoning paradigms. Furthermore, even after these models have been constructed in a given situation, they usually cannot be transplanted without modification to other situations. Usually, they must be particularized or adapted to fit local conditions, constraints, and opportunities. In other words, the models tend to involve situated knowledge when they are constructed originally; and they often must be restated when they are used in other contexts.

• Mathematical models cannot always be expressed at a single level of analysis. Mathematical models tend to be partly embedded in a variety of notation systems (written symbols, spoken language, concrete or graphic representations, or experience-based metaphors) each of that emphasize or facilitate somewhat different aspects of the underlying model. Furthermore, every model ignores (or distorts) some aspects of reality in order to emphasize (or clarify) others. Therefore, a community of partly incompatible (and yet partly overlapping) models may be needed to interpret and explain a given situation. For example, a well known instance of this phenomenon occurs in physics in the case of the wave and particle models of light. But, similar instances occur routinely in nearly every area of applied mathematics – even at elementary levels.

Even in situations that can be described using elementary mathematics, it is often necessary for students to use several models in parallel and in sequence – and to use a series of several modeling cycles at each phase. For example, consider the following 40-minute problem. This particular problem is typical of those we had used in an earlier project about Using Mathematics in Everyday Situations (Lesh & Akerstrom, 1982; Lesh, 1985). It is a problem that was design-ed for use with average ability seventh graders who worked in situations that were as much like realistic everyday experiences as we could make them in a school environment.

Near a school in a Chicago suburb, a research assistant (with a map of Chicago in hand) walks up to a group of 2-3 students and asks, *What's the best way to get from here to O'Hare Airport?*

Note that in the preceding problem, the term 'best' was left undefined because we wanted to go beyond investigating answer-giving abilities to investigate processes involved in problem formulation, information interpretation, and trial solution evaluation. A map was available, but no other suggestions were given about whether best was intended to mean: *shortest, quickest, safest, simplest, least confusing, most convenient, least expensive,* or some other possibility. Also, no suggestions were made about whether a car was available – or whether a bus, taxi, limousine, or train might be needed. However, these kinds of information were available if students requested it.

Let's briefly consider the nature of a typical solution to such a problem. When students first began to work on the O'Hare problem, alternative ways to think about the situation were seldom explicitly recognized or sorted out. Students usually began by simply looking at the map to try to identify a route that was an unconscious mixture of *short/quick/easy/cheap*. Furthermore, they tended to notice only a few items with respect to any given way of thinking about the problem; and, without explicit awareness, they tended to switch back and forth among *short, quick, easy,* or *cheap* types of factors. At the same time, they often made unwarranted assumptions about the problem situation – or imposed unnecessary constraints. For example, without examining their assumptions, they often assumed that (i) a car was available, (ii) paying for parking was not an important consideration, and (iii) leaving the car in a parking lot for an extended period of time presented no difficulties. Gradually however, as the students sorted out and refined several of the preceding interpretations (*short, quick, easy, cheap*), they began to notice more details, and several interpretations began to get sorted out and integrated – so that, by the end of a 40 minute session, answers that were generated tended to have the following characteristics.

- They were detailed (e.g., for any given route, a great many details were noticed related to factors such as time, convenience, or possibilities of traffic jams.)
- They were differentiated (e.g., different kinds of factors were sorted out, and trade-offs were noted -- such as the fact that potentially fast routes were often high risks for accidents or traffic jams.)
- They were integrated (e.g., If you care about factors A, B, and C, then you should choose a way that deals in a combined way with these factors.)
- They were conditional and flexible (e.g., If it's rush hour, then do X; if not, do Y or Z.)
- They were aware of assumptions and possible sources of difficulties or errors (e.g.,If your car is in the lot for a week, you'll pay a lot, and your radio may get stolen.)

It is useful to identify a few additional characteristics of the O'Hare problem and its solutions. For example: (i) No tricks were needed; and, solutions only required straightforward uses of elementary mathematics. So, students rarely confronted the difficulty of being stuck or at a loss for ideas to try. (ii) During 40-minute solution attempts, students seldom spent more that 5-10 minutes engaged in number crunching activities (or even activities that could be characterized by getting from explicit givens to goals). Instead, their most significant activities had to do with figuring out, refining, and re-defining their interpretations of givens and goals. (iii) Final solutions tended to involve coordinating, differentiating, and integrating several (initially unstable) conceptual models concerning time, distance, cost, and so on. (iv) Because the problem involves both too much and not enough information, models had to be constructed so that meaningful patterns could be used to filter and/or simplify the situation and fill in holes, or go beyond information given (Lesh, 1990).

In general, the most important characteristic of problems from the preceding project were that the final result generally involved constructing one or more models. That is, within a given 40-minute problem-solving session, average ability students routinely invented (or extended, or refined) significant mathematical objects by going through a series of modeling cycles, each of which interpreted the givens and goals in distinctive ways.

To further illustrate the preceding points, consider the kinds of solutions that average ability middle-school children generate for the following problem, which is another example from our research about Using Math in Everyday Situations (Lesh & Zawojewski, 1987).

Problem: Fred Findey began teaching here at the high school 10 years ago. He and his new bride rented an apartment at 318 Main Street for $315 per month, and he also bought a new VW Rabbit for $6,200. His starting salary was $16,300 per year.
This year, Fred's sister, Pam, also began teaching at the high school. Pam, too, just got married. In fact, she rented the very same apartment as her brother did 10 years ago, only now the rent was $610 per month. She also bought a new VW Rabbit for $13,700. Using this information, and these *(see items below)* newspapers and catalogues, determine how much do you think Pam should get paid?

Note: Students were given: (i) a calculator; (ii) two Sears [major department store] catalogues (one current, and the other from 10 years ago), and (iii) two newspapers (one current, and the other 10 years old).

For the kinds of reasons described below, we have referred to problems from the preceding project as *concept-eliciting problems,* or as *local conceptual development problems* (Lesh, Post, & Behr, 1989).

• The development of an adequate interpretation of givens and goals tends to be at least as important as the process of stringing together procedures for getting from the initial conditions to the desired state. Therefore, solutions inevitably require students to construct an adequate conceptual model to interpret the problem situations.

• Students initial interpretations of givens and goals tended to be based on interpretations that were: (i) unstable, and (ii) barren and distorted – in the sense described in the preceding section. For example, students early conceptualizations of the problem tended to take into account only a small and biased subset of information, and to be based on a number of unwarranted assumptions.

• Students solutions usually involved between three and ten modeling cycles in which a series of tentative interpretations of the problem had to be progres-sively constructed, sorted out, and tested – and then rejected, modified, refined, or combined with other models.

• Later interpretations of the problem were, in general, based on: (i) more information than earlier interpretations, (ii) richer patterns of information, and (iii) more accurate and precise information. Furthermore, later interpre-tations tended to take into account: (i) explicit analyses of underlying assump-tions, (ii) possible sources of errors, and (iii) conditions that might specific answers might need to be modified.

In other words, significant local conceptual developments tended to occur during 40-minute solutions to our concept-eliciting problems. For example, during 40-minute solutions to the preceding inflation problem, average ability seventh grade students often constructed (locally at least) several foundation-level models related to ratios, proportional reasoning, statistical sampling, and/or weighted averages. That is, solutions tended to involve creating, modifying, or refining several of the most important conceptual models (or reasoning patterns) in elementary mathematics. Furthermore, when the underlying model needed to interpret the situation corresponds to a foundation-level concept that has been studied by developmental psychologists, the sequence of models that students go through during 40-minute solutions tend to correspond to compact versions of the stages that developmental psychologists have observed over time periods of several years. For example, Lesh, Post, & Behr (1989) give details about how the local modeling cycles that students went through for the inflation problem tended to be nearly identical to the stages that were observed over time periods of several years concerning the natural evolution of ratio and proportional reasoning concepts.

The preceding observations suggest that there are important similarities between (i) psychological modeling as it was described in the preceding section of this chapter, and (ii) mathematical modeling as it is understood by professional mathematicians. That is, the processes that children use to construct informal (elementary) models to make sense of their everyday experiences are similar to the processes that mathematicians use to construct formal mathematical models to describe, transform, or predict more complex situations. Furthermore, this section will give an example from elementary mathematics to illustrate similarities between: (i) the processes that both mathematicians and children use to construct local models to interpret particular problem-solving situations, and (ii) the processes that are driving forces behind children's general conceptual *development*. For example:

• To explain general conceptual development, we can use the processes that (both) mathematicians and children use to construct mathematical models to make sense of local problems. In other words, at the level of either elementary mathematics or advanced mathematics, it is productive to think of general conceptual development as being the result of gradually increasing local competence, rather than as being a global manifestation of some general cognitive structure.

• Conversely, to describe some of the most important kinds of problem-solving processes, heuristics, and strategies that are needed to solve problems in particular problem-solving situations, we can use the mechanisms that developmental psychologists have relied upon to explain the development of children's natural conceptual models (which are based on general everyday experiences). In other words, at the level of either elementary or advanced mathematics, the most important types of realistic problem-solving sessions can be interpreted as local conceptual development sessions.

That is, the processes that contributed to conceptual development involved coordination, integration, and differentiation – with the relevant conceptual models being constructed while student passed through the following kinds of recursive refinement-and-elaboration cycles.

• When a model became sufficiently well-coordinated and flexible: (i) new within-model mismatches were detected, and/or (ii) new model-reality mismatches were detected.

• When new mismatches were detected, students had to be extend or specialize their model – with this newly added complexity creating the need coordination to be re-achieved.

• When coordination is re-achieved, the preceding cycle repeats until learners (or problem-solvers) decide that the predictions from their models are good enough to achieve perceived goals!

Cognitive conflicts (or interpretation mismatches) are driving forces creating the need for progressively more powerful and stable models and reasoning patterns ; and, new reasoning patterns are created by coordinating, integrating, differentially, and reorganizing existing models and reasoning patterns. Therefore, if an AI-based tutor aims at facilitating conceptual growth in the preceding kinds of contexts, then it must find ways to confront students with (i) model-reality mismatches, e.g., by encouraging students to make predictions into real situations (or computer-based models or simulations), (ii) between-model mismatches, e.g., by using multiple linked representations to highlight similarities and differences among existing models, as well as highlighting multiple views of individual models, and (iii) within-model mismatches, e.g., by using computer-based microworlds in which targeted conceptual systems are facilitated, and where tools (such as calculators, spread sheets, symbol manipulators, and function plotters) can be used to help students execute low level procedure without getting so caught up in details that they lose the big picture.

Recommendations for AI Tutor Building

The first and perhaps most difficult recommendation for building AI tutors along the lines suggested above is that one must be willing to look less to the program language capabilities and knowledge representational systems of the computer as the design framework for the tutor. Rather, the design must grow out of answers to the following kinds of questions: What is my model of mathematical problem solving? How do I believe that students learn mathematics? Am I willing to forego precision in the computer's model of the student in favor of having the student (using many computer-based tools) construct his or her own model – even if this necessitates dealing with fuzzy or ill-defined products?

Concerning the nature of mathematics: Tutor designers that think of mathematics as simply a list of facts and skills, will tend to think mainly in terms of (i) explaining and demonstrating these facts and skills, and (ii) correcting or debugging students' faulty memories of skills. Rather they should consider molding and shaping students' higher-order reasoning patterns (using the rich problems described above) even if descriptions of these processes are not wholly pre-programmable.

Concerning the role of authority and the AI tutor: Rather than casting the tutor in the role of an external authority about the quality of students' responses, tutor designers should create situations (e.g., provide examples, create tasks, or ask questions) in which students themselves could judge the correctness, usefulness, or adequacy of their own responses. For example, two ways this can be accomplished include: (i) asking students to make predictions about real situations or simulations, or (ii) using alternative representations to help students think about their own thinking, e.g., by unpacking and re-assembling previously unanalyzed models or thought processes. Using the preceding approach, the tutor should follow the students' reasoning far more than the student follows the tutor's reasoning because apparently incorrect steps often led to very productive results in the long run by helping students clarify their understanding of the basic problem situation.

Concerning the role of errors: Procedural errors can often be simply ignored – these errors should be addressed by CAI drills rather than AI tutors. Bugs should be viewed as symptoms of deeper difficulties, rather than being the real source of difficulties. Therefore, in order to avoid getting bogged down in computational distractions tools such as calculators should be encouraged.

Concerning the role of negative feedback: Negative feedback should be avoided not only so as not to hurt students' motivation, but also because the correctness of answers is not really the main concern.

The main goal is to encourage students to explore the quantitative structure of mathematically rich problem situations (Thompson & Lesh, in press), and to provide conceptual mirrors to help them monitor and assess their own plans and processes.

Tutors can avoid negative feedback by focusing on richer and more multifaceted problems, tasks, and questions so that many levels and types of correct answers are possible – depending on factors such as the reasoning patterns that were used, the assumptions that were recognized, or the type of information that was ignored or taken into account. In this way, some aspect of each student's responses are nearly always correct. Then, tutors can focus on building on these positive aspects of students' responses — rather than being preoccupied with the elimination of negative aspects. Because the goal is to explore the ideal conceptual terrain corresponding to underlying structural characteristics of the given problem situations, errors are not necessarily things to be avoided. In fact, skillful tutors should sometimes torpedo students by setting them up to encounter cognitive conflicts even in cases where the students original response to a problem had been correct. For example, by varying problems along a number of structurally interesting dimensions, cognitive mismatches are sometimes induced to help students recognize the need for conceptual re-organizations (see in particular Krutetskii, 1976).

Concerning the selection of problem sets: Tutors should not be very concerned about covering all problems within given problem sets. Rather, a small number of rich problems should be fully explored. For example, after a student had responded (correctly or incorrectly) to an original problem, tutors sometimes should change the values of the givens and then asked for a corresponding modification in the answers. Or, some of the givens and goals can be reversed, and students asked to make predictions about the consequences. Or, tutors can ask for alternative descriptions of the basic problem situation – using diagrams, concrete materials, or other representations. Well structured questions (like the examples above) can be interpreted at a variety of levels. That is, the problems tend to be self-adjusting in difficulty because students interpret them in a variety of ways and at several levels (again, see Krutetskii, 1976).

To compensate for the preceding fuzziness in problem interpretations for individual problems, tutors have to be much more clever about encouraging students to investigate similarities and differences among related clusters (or sequences) of problems. In other words, the same sort of instructional dilemma arises that tends to arise when Dienes-style concrete embodiments are used in mathematics laboratory forms of instruction (Dienes, 1957). The kernel of the dilemma is that, even though individual tasks are carefully structured to encourage the construction of specially targeted conceptual models, it is also important for students' responses to be open-ended. But, how can instructional focus be given to cluster of somewhat unfocused tasks? How can the tutor maximize the chances that students will focus on built in structures of tasks? The techniques that Dienes recommends, and the techniques that teachers tended to emphasize toward the end of our project, involve going beyond randomly organized strings of isolated embodiments (or problems, or tasks) to focus on multiple embodiments – and on similarities and differences among structurally related tasks.

Past publications have given details about how Dienes' instructional principles can be implemented in computer-based learning environments (Lesh, Post, & Behr; 1987). For example, we have described some important ways to use interactive graphics in a manner similar to the way Dienes uses concrete manipulatives. Now, the claim is being made that concept-eliciting problems can also be used in a manner similar to the way concrete manipulatives have been used in the past. In fact, using concept-eliciting problems in this way is an essential characteristic of the kind of AI-based tutor that we favor.

Concerning the intelligent generation of hints, helps, or next steps: Tutors should help students see an overall map of the conceptual terrain that they are expected to traverse; and, encourage the use of graphics and diagrams, or technological tools (such as graphics-linked calculators) to: (i) carry out elementary-but-tedious tasks that might distract the student's attention from more important activities, (ii) become familiar with linked representations to provide conceptual mirrors (Schwartz, 1989) so that students could view their own models, activities, and results from multiple perspectives, or (iii) encourage them to maintain graphic ledger sheets (Collins, Brown, & Newman,1990) to display map-like records of processes.

Extensive follow-up questions can and should be generated for a problem that was perceived to be particularly rich, regardless of whether a students original answers were correct or incorrect. Examples of such probes include: (i) Here is a situation, describe it (using graphs, materials, symbols, etc.). (ii) Now, use your description to make a prediction (e.g., about a simulation, or a realistic situation). (iii) Can you refine your description in the light of your prediction? (iv) Here is a similar situation, describe a problem like the one you just solved. (v) Apply your previous solution to the new problem, or modify it in appropriate ways. (vi) Complete this diagram of the problem. (vii) Simplify the problem statement by crossing out all of the irrelevant words. (viii) Identify the givens. (ix) Here is a problem that is similar to the one you just finished, and here is a diagram of the solution steps you used. Show how to solve the new problem by inserting its givens into appropriate boxes in the diagram (see Polya, 1957). In other words, sequences of challenges can be given in which students can judge the quality of their own answers. Next steps are then based more on students' personal preferences (within a predetermined conceptual landscape) than on the tutor's preferences. Therefore, the tutor's duties focus on keeping students informed about options that are available to them, and on facilitating progress – rather than on limiting (or dictating) students choices using fine grained hints, helps, or follow-ups.

Concerning expert-novice characterizations of knowledge: A students state of knowledge is seldom similar to a fixed (and perhaps flawed) set of rules in a computer program. Instead, each students knowledge is situated, and that it is often relatively local and unstable. Therefore: (i) a student's errors or biases in one context should not necessarily assumed to transfer to other situations, (ii) a student's capabilities in any given task would be strongly influenced by what Vygotsky (1962) refers to as a zone of proximal development, and (iii) most ideas can be assumed to be at intermediate levels of development – rather than being 100% mastered or not mastered.

One of the best ways to get a valid snap shot of a students current conceptual abilities is to try to mold and shape the conceptual models and reasoning patterns on which these abilities are based. In other words, one of the best ways to clarify the nature of a students current state of knowledge is to try to induce changes in that state.

Concerning the role of preplanned activities: Knowledge about a students models and reasoning patterns is gained primarily by observing the student's reactions to instructional strategies which are not created at run time (during an ongoing tutoring session), but are instead planned in advance – based on the tutor designer's sophisticated understandings about a relevant *idealized model of potential student knowledge.* Tasks should be based primarily on extremely rich conceptions about what it means to understand the relevant terrain's of knowledge – rather than being based on a detailed characterization of a particular student's knowledge.

Concerning the conception of instruction: Conceptions of instruction need to go beyond constructing new conceptual models to recognize the need to reject old ones – or to make a conceptual shifts from old models to new models. For example, in science education, helping students recognize the need for a new idea (such as inertia) is often the heart of what is required to understand the idea. Often, the goal of instruction is not so much to disseminate pearls of wisdom as it is to disseminate grains of sand that will stimulate teachers to create their own pearls. Therefore, tutors can aim at confronting students with cognitive conflicts (i.e., creating model-reality mis-matches, or within-model mismatches) in order to confront students of the types needed to encourage the construction of new models (or the rejection of old models).

Conceptual models develop as they are gradually embedded in progressively more powerful systems of ideas; and the meanings of conceptual models gradually evolve along a variety of dimensions including: concrete-to-abstract, simple-to-complex, particular-to-general, external-to-internal, and intuitions-to-formalizations. So, again, models develop by integrating, differentiating, and modifying existing systematic wholes – not simply by constructing wholes out of parts.

Students must go beyond thinking *with* conceptual models to also think *about* them: (i) by embedding the relevant underlying models within concrete/graphic embodiments (or powerful prototypic situations) and by unpacking and reassembling the models in a variety of ways, and for a variety of purposes, (ii) by embedding the models within a powerful notation system, and then exploring consequences that this meaningful system generates, and (iii) by evaluating underlying assumptions of the models, and by investigating the goodness of fit when the models are used to describe (and make predictions about) realistic problem-solving/decision-making situations.

If the students goes beyond thinking with the system, and begins to consciously think about the system, then the probability goes up that a system will be used, and that it will be used in a well coordinated fashion. On the other hand, if a conceptual system-as-a-whole doesn't function as a systematic whole, then many of the most important properties it should describe tend to be invisible to the student. For example, such students often become so involved with materials (and with isolated actions) that they have difficulty noticing patterns that are being used to organize and manipulate the materials. They often have difficulties seeing the forest because of the trees.

To summarize, the practical activities in tutor designing should be: (i) to identify concept-eliciting problems whose solutions tend to involve the construction of these targeted models, (ii) to use feedback's, helps, and follow-up questions aimed at facilitating the preceding mechanisms for promoting conceptual adaptation, and (iii) to go beyond using isolated problems to focus on clusters of structurally similar tasks, and/or to go beyond isolated representations of a given problem to juxtapose multiple linked representations of the problem and to focus the students' attention on structural similarities and differences in these related contexts. That is, the final activity involves going beyond thinking with a given conceptual model to also think about the underlying interpretation framework; they involve making the reasoning pattern itself an object of study ; and, they involve going beyond thinking within a given notation system, to also explore similarities and differences among alternative graphic, written symbolic, or spoken descriptions of the problem.

Conclusion

In conclusion, current cognitive and mathematical theories indicate that we must find ways to produce good practitioners without being able to reduce our conception of expertise to a short list of easy-to-execute rules. There is no single pre-programmable definition of a good tutor partly because good tutors are key elements within the system in which they are attempting to optimize results. Their design involves pre-planning structurally rich environments similar to those Dienes (1957) and others have used to create multiple embodiment activities for students; and, facilitating the construction of specially targeted cognitive systems. The relevant approaches appear to be similar to: (i) techniques that excellent coaches use to help athletes construct progressively more complex, flexible, and well coordinated systems of physical activities, (ii) processes that Piaget and other developmental psychologists have used to explain the evolution of general cognitive structures, and (iii) techniques that students use to extend, reorganize, and refine their own thinking during realistic everyday problem-solving situations.

References

American Association for the Advancement of Sciences Project 2061 Science for all Americans. Washington, D.C.: AAAS, 1989.

American Association for the Advancement of Science. Project 2061 (1989). What Science is most worth knowing? Washington, D.C.

Bunderson, V. A. (1983) The diagnosis and remediation of cognitive bugs in arithmetic. (with J.B. Olsen). Final Report, NSF Contract #SED 80-12500. WICAT Education Institute Technical Report, September.

Carey, S. (1988). Reorganization of knowledge in the course of acquisition. In S. Strauss (Ed.), Ontogeny, Phylogeny, and historical development. Norwood, NJ: Ablex.

Clement, J. (1982a). Students preconceptions in introductory mechanics. American Journal of Physics, 50, 66-71.

Collins, A.; Brown, J. S. ; & Newman, S. E. (1990). cognitive apprenticeship: Teaching the craft of reading, writing, and mathematics. In L. B. Resnick (Ed.) Cognition and instruction: Issues and agendas. Hillsdale, NJ: Lawrence Erlbaum Associates.

Cronbach, L. J., & Snow, R. E. (1977). Aptitudes and instructional methods. New York: Irvington.

De Kleer, J. & Brown, J.S. (1980). Mental models of physical mechanisms and their acquisition. In J.R. Anderson (Ed.), Cognitive skills and their acquisition. (pp. 285-309). Hillsdale, NJ: Erlbaum.

Dienes, Z. (1960). Building up mathematics. London: Hutchinson Educational Ltd.

diSessa, A. (1989) Knowledge in pieces. In G. Gorman & P Pufall (Eds) Constructivism in the Computer Age. Hillsdale, NJ: Lawrence Erlbaum Associates.

diSessa. S. (1987) Toward an epistemology of physics. Cognitive Science Technical Report. Berkeley, CA: Institute for Cognitive Science.

Krutetskii, V.A. The psychology of mathematical abilities in school children. Chicago: University of Chicago Press, 1976.

Lesh, R. & Kelly, A. (in press) The development of tutoring capabilities in middle school teachers. Journal for Research in Mathematics Education. Reston, VA: NCTM.

Lesh, R., & Akerstrom, M. (1982). Applied problem solving: Priorities for mathematics education research. In F. Lester & J. Garofalo (Eds.), Mathematical Problem Solving: Issues in Research (pp. 117-129). Philadelphia: Franklin Institute Press.

Lesh, R. (1985). Processes, skills and abilities needed to use mathematics in everyday situations. Education and Urban Society, 17 (4), 439-446.

Lesh, R. (1989). Computer-based assessment of higher-order understandings and processes in elementary mathematics. In J. Kulm (Ed.) Assessing Higher Order Thinking in Mathematics. AAAS: Washington, DC

Lesh, R., Post, T. & Behr, M. (1989) Proportional reasoning. In M. Behr & J. Hiebert (Eds.) Number Concepts and Operations in the Middle Grades. Reston, VA: NCTM.

Lesh, R., Post, T., & Behr, M. (1987) Dienes revisited: Multiple embodiments in computer environments. In I. Wirszup & R. Streit (Eds.), Developments in School Mathematics Education Around the World. Reston, VA: NCTM.

Lesh, R. & Zawojewski, J. (1987). Problem solving. In T. Post (Ed.), Teaching Mathematics in Grades K-8 : Research-based Methods. Boston: Allyn & Bacon.

Lesh, R. (1990). Computer-based assessment of higher-order understandings and processes in elementary mathematics. In J. Kulm (Ed.) Assessing Higher Order Thinking in Mathematics. AAAS: Washington, DC

Mathematical Sciences Education Board, National Research Council. Reshaping School Mathematics. Washington, D.C., National Academy Press, 1990)

McCloskey, M. (1983). Naive theories of motion. In D. Gentner & A. Stevens (Eds.) Mental models. Hillsdale, NJ: Erlbaum.

McCloskey, M. (1983) Intuitive Physics. Scientific American, 284, 222.

Minstrell, J. (1982). Conceptual development research in the natural setting of a secondary school science classroom. In M. B. Rowe (Ed.) Education for the 80's: Science. National Education Association.

Naisbitt & Aburdene, 1990 National Council of Teachers of Mathematics' Curriculum and Evaluation Standards for School Mathematics. Reston, VA: NCTM.

National Council of Teachers of Mathematics. (1989). Curriculum and evaluation standards for school mathematics. Reston, VA: NCTM.

Polya, G. (1957). How to solve it: A new aspect of mathematical method. (2nd. ed.). Princeton, NJ: Princeton University Press. (Original work published 1945).

Polya, G. (1984). With, or without, motivation. In Gian-Carlos Rots (Ed.). George Polya: Collected papers, Vol. IV. (pp. 496-503). Cambridge, MA: MIT Press.

Schwartz, J. L. (1989). Intellectual mirrors: A step in the direction of making schools knowledge-making places. Harvard Educational Review.

Sleeman, D., Kelly, A. E., Martinak, R., Ward, R., & Moore, J. (1989). Studies of diagnosis and remediation with high school algebra students. Cognitive Science, 13 (4), 551-568.

Thompson, P. & Lesh, R. (in press) Quantitative Structures and Procedural Structures in Mathematical Problem Solving. In R. Lesh & S. Lamon. Assessing Deeper and Higher-Order Understandings in Elementary Mathematics. ETS Princeton, NJ.

Thompson, P. W. (1989). Artificial intelligence, advanced technology, and learning and teaching algebra. In S. Wagner & C. Kieran (Ed.), Research issues in the learning and teaching of algebra. (pp. 135-161). Reston, VA: NCTM.

Winograd, T., & Flores, F. (1985). Understanding computers and cognition: A new foundation for design. Norwood, NJ: Ablex.

Appendix

A Ten-Week Study of Twenty Human Tutors

The study focused on a ten-week project in which real human tutors provided the intelligence behind a computer-based tutoring system. Then, for teachers who were at various levels of expertise, our goal was to investigate the understandings, assumptions, and procedures that were really used in tutoring – and to observe the evolution of these tutoring abilities over time, as tutoring effectiveness improved.

The tutors: The study involved two groups of typical middle school teachers, with ten teachers in each group. Each week during a ten weeks period, each teacher participated in two 60-minute tutoring sessions which were held on either Tuesday and Thursday, or Monday and Wednesday. Then, on Friday of each week, each group of ten teachers met independently to give brief written-and-oral reports about their two tutoring experiences for the week. Also, during these Friday sessions, the group discussed the pros and cons of various tutoring approaches – and refined their collective conceptions about the nature of good tutoring techniques, good problems, good follow-up questions or feedbacks, and good helps or hints. The main goal of the study was to investigate the ways in which our teachers' tutoring capabilities would evolve during the preceding ten week period. In particular, we were interested in changes that might occur concerning: (i) the problems and follow-up questions that would be emphasized, (ii) the kinds of answer-checking, feedbacks, hints, and helps that would be used, and (iii) the rules that would be used to determine when the preceding tutor inputs would be initiated.

The students: Forty students participated in the study (8 fifth graders, 12 sixth graders, 8 seventh graders, and 12 ninth graders), with each student participating in one tutoring session per week for five consecutive weeks. That is, each teacher selected four students from his or her class, with selections being based on the teacher's' judgements about children who might profit most from one-to-one tutoring. Therefore, nearly all of the students who were selected were viewed as being average or below average – but as having potential to improve!

The laboratory setting: During the tutoring sessions, the students were told that they were interacting with a smart computer ; and, indeed, they were! How-ever, the main components of the computer's intelligence were provided by their teacher, who was able to view the student's activities from behind a one-way mirror – or using closed circuit TV monitors, one focused on the students face, and one focused on the students pencil-and-paper activities. In other words, teacher-student pairs worked together on two linked PCs in which special windows on the students screen enabled teachers to communicate with students using a local E-mail system to transmit: (i) brief messages which were written by the teacher prior to the on-line tutoring session, or (ii) simple graphics which the teacher could create at run-time. In addition to being able to view the students' behaviors through a one-way mirror, or using closed circuit TV monitors, the teacher was also able to view two computer screens: (i) one showed an exact duplicate of the student's screen, and (ii) the other showed a menu of problems, feedbacks, helps, and hints which had been by the teacher prior to the on-line tutoring session (as shown in Figure 1).

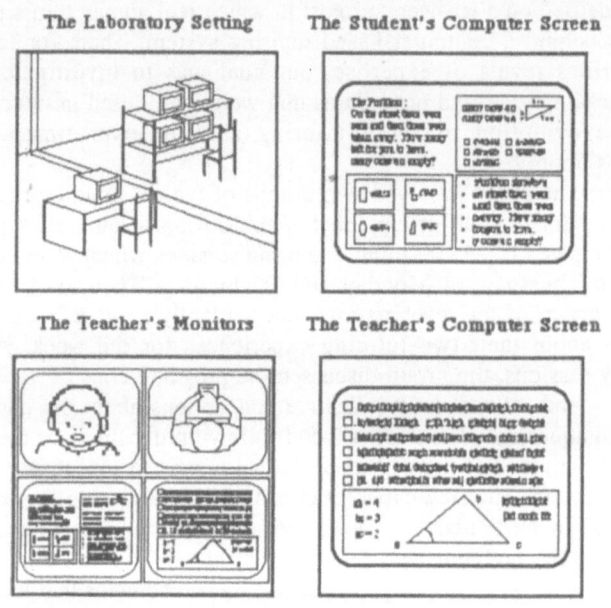

Figure 1

The teacher's pre-tutoring assignments: On the day before each tutoring episode was to occur, each teacher was given one of the following two assignments designed to help them prepare for the up-coming tutoring session. The assignment was somewhat different on odd-numbered weeks and on even-numbered weeks.

• During odd-numbered weeks, on the day before the tutoring session, each teacher was given a scored ten-item worksheet that the student had completed – and on which the student needed tutoring. Then, the teacher was asked to write a set of problems which they could use during the up-coming tutoring session, with an aim of helping students prepare for a post-tutoring test (similar to the pre-tutoring worksheet) that would be given to the student after the tutoring session.

• On even-numbered weeks, the teachers were given the problems which would be used during the tutoring session. That is, each teacher was given a ten-item problem set which included problems about a single topic area (such as similar figures or proportional reasoning). These problems were explicitly constructed to include a variety of types ranging from routine word problems to more innovative problems such as those in which several levels or types of correct answers, or those which included with too much and/or not enough information, or those which were designed to capitalize on the fact that calculators were available.

Also, in both of the preceding two situations, the teachers used a take-home PC to write feedbacks, follow-up questions, hints, and helps for each problem that would be used during the tutoring session. The goal was to anticipate which kinds of questions, follow-ups, feedbacks, hints, and helps might be most useful to students. Then, prior to the tutoring session, the preceding teacher-generated items were up-loaded into an easily accessible menu of options on the computer system that would be used during the tutoring session. In this way, during the tutoring session, the teachers were able to simply click on problems, feedbacks, follow-ups, hints, and helps, in order to send them as e-mailed messages to the student.

The Friday discussion groups: On Friday of each week, each group of ten teachers met together independently to give brief written-and-oral reports about their two tutoring experiences for the week, and to discuss the usefulness and productivity of various types of problems, follow-ups, feedbacks, helps, and hints. We also introduced examples of new problem types and *techniques that teachers at another site have used* which we wanted our teachers to consider. Finally, we spent a small amount of time describing students' solutions to realistic everyday problems such as the inflation problem and the getting to the airport problem in which finding adequate ways to think about the problem involves constructing new (to the student) conceptual models for interpreting givens, goals, and possible solution paths. In general, we wanted our teachers to at least consider the possibility of using tutoring techniques that focus on processes similar to those that students themselves use to improve their own concep-tualizations of such problems.

The post-session debriefing sessions: On even-numbered weeks, a tape recorded post-tutoring interview was conducted with half of the teacher. In these sessions, teachers were asked to recall and describe the first tutoring session that they had conducted during the preceding odd-numbered week. They were also were shown a problem set similar to the one that had been used during the tutoring session, and were asked to predict how the student performed (in terms of accuracy, insights, and errors) when the given problem set was used as a post test following the tutoring session. (note: Teachers were not shown actual results from the post-tutoring tests until after the post-tutoring interviews had been completed, and predictions had been made.) Our goal was to investigate possible relationships between: (i) the quality of tutoring sessions and (ii) the complete-ness and accuracy of recalled and predicted information.

The tutoring quality evaluations: On even numbered weeks, information about each teacher's tutoring effectiveness was gathered from four sources: (i) improvements, or lack of improvements, noted in pre-to-post-session assessments of students capabilities, (ii) teachers' self assessments of their own tutoring effectiveness for the week – compared with the effectiveness of peers, and compared with their own effectiveness during previous weeks, (iii) teachers' retrospective re-evaluations of tutoring effectiveness during previous weeks, (iv) teachers' assessments of their peers' tutoring effectiveness for the week, (v) evaluations by three experts on our project staff concerning each teacher's tutoring effectiveness for the week, and (vi) experts' retrospective re-evaluations of tutoring effectiveness during previous weeks. These latter assessments were based on: (i) analyses of print-outs and oral reports describing the types of problems, follow-ups, hints, and helps that were used, (ii) observations about the insightful use of these teacher inputs during tutoring sessions, and (iii) observations of insights gained by students during the tutoring sessions.

An Overview of Results

General results from the preceding study included the following.

(i) For a given tutoring session, pooled evaluations of tutoring quality were highly correlated with the teacher's ability to recall rich, detailed, and accurate information during post-session debriefing interviews; it was highly correlated with the depth and sophistication of descriptions that teachers were able to give during post-session interviews based videotapes of selected sessions; and, it was highly correlated with the teacher's ability to make accurate predictions about how the student would perform during post-session tests.

(ii) The quality of tutoring increased significantly during the ten-week project – in which quality assessments were based on a combination of students' learning gains, teachers' self-assessments, teachers' peer-assessments, and assessments by expert tutors on the project's staff.

(iii) During sessions when the tutoring was least effective, the teachers generally appeared to operate (implicitly) using rules of the type that have characterized traditional types of AI-based tutors; whereas, during sessions when tutoring was most effective, the preceding principles tended to be explicitly rejected – and were replaced by principles based on the constructivist perspective described briefly in the preceding chapter.

The Influence of Interactive Tools in Geometry Learning

Heinz Schumann

Fachbereich Mathematik, Pädagogische Hochschule Weingarten
Kirchplatz 2, D-88250 Weingarten, Germany

"As designers, it is our duty to develop systems and instructional materials that aid users to develop more coherent, usable mental models. As teachers, it is our duty to develop conceptual models that will aid the learner to develop adequate and appropriate mental models..." (Donald A. Norman 1983)

1. Introduction

An essential procedural aim in secondary education is the learning to solve problems. Problem solving its to be seen as the most complex form of learning, it can include concept learning and rule learning. We hope and believe that the abilities and skills developed through problem solving in secondary schools can be transfered to extra school activities. In geometry teaching we can differentiate between the following typical kinds or ideal types of problems: construction problems, calculation problems, theorem finding problems, proving problems ...

Common to these problems are above all the heuristic strategies and the iconic representation, which are used in solving them. Unfortunately in the context of school geometry only solvable problems are treated. The most concrete for students are construction problems. The measurement of geometric objects leads into calculation problems. In the calculation problems the geometric figure is connected with number operations on the level of the numerals or number variables. The solution of construction problems form a basis for the finding theorems, which can be viewed as a new problem – problems or problem solutions generate new problems. The subject of proving problems is the search for the relative verification of theorems, which can be realized on various intellectual levels. We must distinguish between more subjective and more objective solutions of proof finding problems; at this point the whole tension between the individual learner and conventional geometry manifests itself. Global tasks of the teacher in the teaching of problem solving are:
- a suitable selection and ordering of the problems to be solved according to the abilities of the learning groups in question,
- the motivation of learners to solve these problems,
- the supporting of the learner in the understanding analysis of problems and in the carrying out and evaluating of the solutions.

Now there are some computer tools, which can supplementary support the learner in some parts of the process of special geometric problem solving, in order to compensate the weaknesses of traditional instruments, media and methods. We are convinced that the use of computers in the form in Intelligent Tutoring Systems (ITS) can improve the learning and teaching of problem solving by a more effective adaptation to learner behavior than has previously been the case with the use of computers. Subsequently we are concentrating on interactive computer-tools as an aid in the solving of construction, calculation and theorem finding problems in secondary school geometry. The reason for this is that the design of these tools is an essential prerequisite for the development of all modules of an corresponding ITS. Without conceptional reflections on such tools the danger of arbritrarily eclectic conceptions of ITS exists. We must confine ourselves to 2D graphics systems because 3D graphics systems are still underdeveloped.

2. Requirements for Interactive 2D Graphics Systems

Which requirements must be fulfilled by a computer tool, in order to be used in a solving environment for construction, calculation undo theorem finding problems. We can distinguish between the interdependent requirements.

2.1 Geometric Requirements

Basic tool is an interactive graphics system for geometric constructions in school geometry. For the following geometric requirement can be assumed: The preserving of traditional instruments for the following reasons: their practical significance, their mathematical and more general their cultural significance in historical context, the continuity of geometry in the school curriculum, the unity of tactile and visual experiences in geometry learning, the safeguarding of a worldwide standard of (non-verbal) communication, the excessive effort involved in providing hardware and software. Therefore the use of computers is supplementary.

First general requirement (conservative requirement)

The process of construction/calculation by a graphics/calculation system must be possible analogically as well as with the traditional instruments, i.e., individual steps of construction must correspond to commands aiming at the same effect (as far as the sensorial possibilities of the system does allow it).

1st special requirement

Performance of straight-edge and compass-constructions concerning the constructions by straight-edge and compass, the graphics system should allow the user to perform the following actions:

The drawing of loci (straight lines and circles):
- – To draw a straight line through two given points,
- – To draw a circle around a given center and through a circular point.

The fixing of further construction points:
- – Intersection points of constructed loci,
- – Selection of points anywhere in the plane, especially on constructed loci.

2nd special requirement

Concerning the construction of geometric figures in secondary school geometry a graphics system should allow the generation of line segments, half lines (rays), angles (which type?), arcs.

3rd special requirement

A graphics system should have a denoting function for geometric objects with the following capacities: flexible positioning, denotation with the usual geometrical terms, no compulsion to denote (automatic positioning of denotation can lead to overlapping or erasure).

4th special requirement

Concerning the use of protractor and ruler, a graphics system should allow the measurement of angles and line segments.

In this context we should point out the inconsistency of geometric denoting in school geometry, which does not distinguish between the object and its quality, for example measure. Another inconsistency is the different representation of figure concepts; for example, is the altitude of a triangle a line, a half line or a line segment?

5th special requirement

In order to solve calculation problems we need a symbolization of line segments and angles in order to have numerical operations carried out with the measured data of these objects using the system. Up till now the requirements have oriented themselves around the compatibility to traditional school geometry, in particular to the compatibility of the solving process. The mere simulation of constructions, measurements and calculations, which can be carried out using the traditional instruments is not sufficiently enough to legitimize the interactive computer use in geometry teaching and learning. Further requirements must be fulfilled in order to open up new possibilities in geometry learning with the aim of a better compensating for the weaknesses of the traditional media.

Second General requirement (perspective requirement)

In addition to the possibility of simulating feasible constructions, measurements and calculations with conventional tools, a graphics system should have the following additional options:

- Execution of basic constructions as independent commands, or rather menu items,
- Definition and application of macro-constructions (for individual increase of the constructional repertoire, etc.),
- Variation of configurations according to the position, extension, etc. (widening of the inductive basis for gaining recognition),
- Dynamic measurement of line segments, angles and (polygonal) areas,
- Generation of calculation terms based on the measured data,
- Production of local lines,
- Repetition of constructional courses,
- Application of further functions of the system for manipulating, editing and managing constructions, and so on.

2.2 Software-Ergonomic Requirements

An important prerequisite for acceptance and efficiency of computer tools for students is the fulfillment of software-ergonomic requirements, which is already an expected standard for professional software-tools. The graphics system for geometric construction/calculation should be student-friendly'. Are the following *software-ergonomic* principles realized?

Principle of the appropriateness of problems: The solution of construction/ calculation problems and the discovery of geometric statements by the student is to be supported without this procedure being unreasonably hampered by specific characteristics of the system (e.g., through interactive communication).

Principle of the conformity of expectations: The interaction with the system should correspond to those expectations of the student, which (s)he already gained from experiences with working processes upon construction/calculation without and with the computer (see also general requirement 1).

Principle of distinctness: The system should present itself to the student as 'distinct', this concerns the temporal and local organization of graphic and verbal information on the screen, their coding and perceptibility (screen design).

Principle of transparency: The system should be 'transparent' for the student; this refers to the command structure, to the depth and width' of the menu tree, to the messages of the system for the student, etc.

Principe of self-explanation: Global and local aids as well as explanations (upon loss of orientation, unknown inquiry, wrong input, exceeding the system's limits of capacity, exploration of the system) in a form comprehensible for the student is realized automatically or they can be called respectively.

Principle of flexibility: Extensive possibilities of adaptation to the student's level of knowledge (for ex. variable menus). The mastery of the system, increasing with the handling of the system by the student has to be taken into account (e.g., through self-definable macros for drawing and construction). The student should be able to interrupt the operation procedure at any time and, after having used another part of the system, to continue it at that position where (s)he interrupted.

Principle of error treatment: Despite input errors it should be possible for the student to come to the desired constructional result or result of calculation. The student is informed about the cause of the error for the purpose of improving it, which is only possible to a limited extend without an (intelligent) tutorial component. The system excuses the student for the constructional error made (at last) by calling an undo function.

There are three basic design possibilities of the user surface of a 2D graphics system:
- Command-driven,
- Menu-driven,
- Desktop-oriented.

Especially desktop-oriented systems, for which the mouse as an adequate input device is a standard, distinguish themselves from the other systems by their high interactiveness, learnability, and user-friendliness.

2.3 Educational Requirements

General educational principles for the evaluation of a learning environment, which places emphasis on the independently acting acquisition of theoretical concepts and structure (like problem solving in geometry) can be gleaned from the article "Some principles for the design of clarifying educational environments" (Moore and Anderson 1969). These principles are: perspectives principle, autotelic principle, productive principle, personalization principle.

"*1. Perspectives Principle:* One environment is more conducive to learning than another if it both permits and facilitates the taking of more perspectives toward whatever is to be learned."

Here the following perspectives must be distinguished: the perspective of the person(s) acting, the perspective of the person(s) reacting, the reciprocal perspective (e.g., the teacher's perspective), the expert's perspective.

"*2. Autotelic Principle:* One environment is more conducive to learning than another if the activities carried on within it are more autotelic."

An activity is called "autotelic" if it contains its own targets and motivational sources, i.e., if it can be performed for its own sake.

"*3. Productive Principle:* One environment is more conducive to learning than another if what ist to be learned within it is more productive."

A cultural object, especially a learning subject, is the more productive the more extrapolations and application it allows. The quality is dependent to a large extent on the logical structure of the subject.

"*4. Personalization Principle:* One environment is more conducive to learning than another if it: (l) is more responsive to the learner's activities, and (2) permits and facilitates the learner's taking a more reflexive view of himself as a learner."

Condition (l) should express that on the one hand the learner can himself determine to a large extend what happens through his own activities, yet on the other hand is motivated to discover interesting connections.

2.4 Three Examples of Contrasting Nature

Firstly, there is no interactive graphics system which fulfils all the previously stated requirements. We will take three 2D graphics systems as examples, which could serve as a basis for an ITS for geometric constructions, calculations, and theorem finding in planimetry, and which create rather different learning environments:
- Geometric Supposer (with the conventional modules Triangles, Quadrilaterals, Circles) – a first pioneer work!
- GEOLOG
- CABRI-Géomètre

The Geometric Supposer is menu-driven; Geolog is command-driven (in the Atari version optionally menu-driven or command-driven); the CABRI-Geometre is desktop-oriented. The Geometric Supposer written in BASIC barely fulfills the geometric requirements for compatibility of the constructing process. The carrying out of geometric measurements and calculations is well conceived but only described by point-demotions. An automatic transfer of constructions is possible, also a repetition-mode, a scaling function and options for basic constructions exist. Other possibilities according to the second general requirement are not realized in the up to date MS-DOS version. From a software-ergonomic point of view one is aware that the Geometric Supposer is developed on a 60-kilobyte machine. A drawing screen adjustment facility should at least be possible from within the system and so on. The educational requirements are only partially realizable an account of the low software-ergonomic standard. The structuring in separate geometric modules encourages the student to think in narrow "geometric drawers".

Theorem finding with the Geometric Supposer: An empirical investigation of theorem finding processes with the Geometric Supposer as a tool was carried out by Michal Yerushalmy (1986), entitled "Introduction and generalization: an experiment in teaching and learning high school geometry" (the phrase "in the Geometric Supposer environment" should now be added to this title). Anybody who intends to carry out a comparative investigation of traditional and computerized learning environments should study this work. The work is particularly valuable on account of its comprehensive diagnostic results which have significance in the study of mental models. (However the experimental design is problematic and the sample is small). When interpreting the investigation results of learning processes in computerized environments we usually encounter the following problem: How dependent are these results on the learning environment by the non-standardized software in question? The NCTM book *How to use conjecturing and microcomputers to teach geometry* (Chazan and Houd 1989) is based totally on the Geometric Supposer. This title unjustifiably claims generality.

GEOLOG (from Geometry and Logic; written in Prolog) nearly fulfills all requirements of solving process compatibility. Angles as referable graphics objects do not exist. However there is a compulsion to denote all basic objects, which is unavoidable in the command-driven systems. The denotations are automatically positioned as in the Geometric Supposer. Essential requirements are fulfilled which go beyond the simulation of activities with traditional instruments, but Prolog is not suitable for designing interactive graphics. Distinguishing between procedure definition and macro-definition is difficult for the naive user. The software-ergonomic standard ist still low. On account of the formal-language input it can be assumed that GEOLOG is educationally effective for only a restricted number of students. For the development of a prototypical expert systems or ITS this kind of restrictive input has certain advantages for system-internal representation.

CABRI-Géomètre (from Cahier Brouillon Informatique pour l'apprentissage de la géométrie) is an entirely new type of tool which possesses certain CAD qualities. Its software-ergonomic qualities should be emphasized. The above mentioned educational principles could be definitely realized by it. The second general requirement is nearly fulfilled; but CABRI-Géomètre is in its present state of development (version 2.0) a pure construction–undo measurement tool, which does allow geometric calculation but only in indirect manner. In the final part of this article we shall look at CABRI-Geometre by describing its constructive facilities in a survey.

3. CABRI-Géomètre Environment for Planimetric Learning

New aspects for learning geometry arise mainly with:
- the inductive acquisition of theorems and formation of concepts,
- the measurement of line segments and angles,
- the construction of local lines,
- the definition and application of macro-constructions,
- the solution of planimetric construction problems, etc.

3.0 The Drag Mode

An essential prerequisite for almost all mentioned interactive constructional possibilities which go beyond the simulation of compass, ruler and geotriangle constructions, is the drag mode: With the drag cursor (in the following figures symbolized by a grasping hand) we can freely move the basic objects of a construction, these are those determining the points, straight lines, or circles of the construction. In doing so, the configuration is transformed. The transformation given by the drag mode is true to straight line and circle (see Fig. 1.1– 1.3; variations of a triangle with mid-perpendicular and circumcircle through dragging a corner of the triangle; Fig. 1.4 is the borderline case). With the drag

mode transformation (according to the system's presetting), the following relations are generally invariant:
- parallelism (see example in Fig. 2)
- orthogonality (see example in Fig. 1)
- partial relation (see example in Fig. 3)
- symmetry of point and straight line (see examples in Figs. 4 and 5)

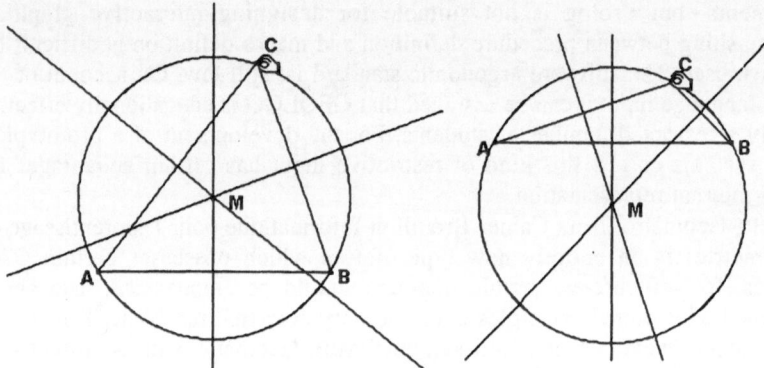

Fig. 1.1–1.2

The relation of incidence does not remain in the event of intersections or points of contact being lost by dragging objects. The drag mode might affect polygons, depending on the basic objects given, which then determine the corresponding polygon totally or partially as: congruent transformation in same orientation, equiform transformation in same orientation, or (local) axis affine transformation (see Schumann 1990/2).

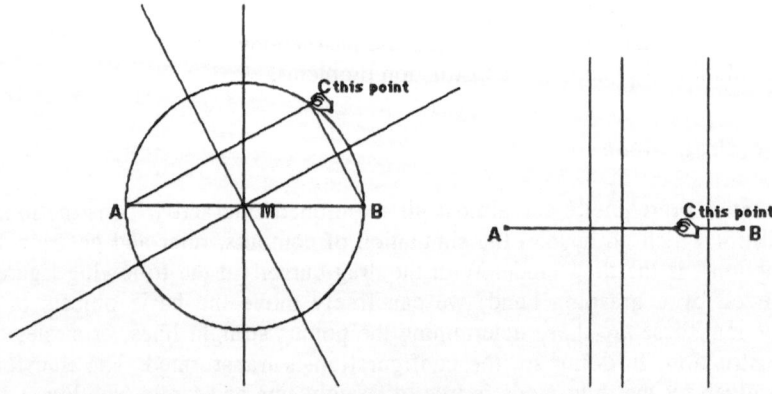

Fig. 1.3–1.4

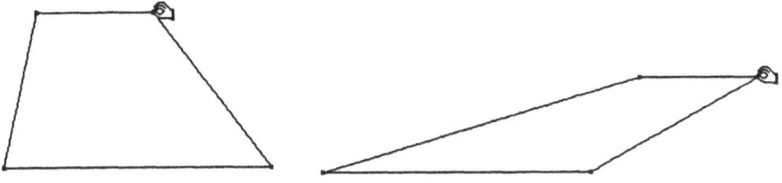

Fig. 2.1–2.2

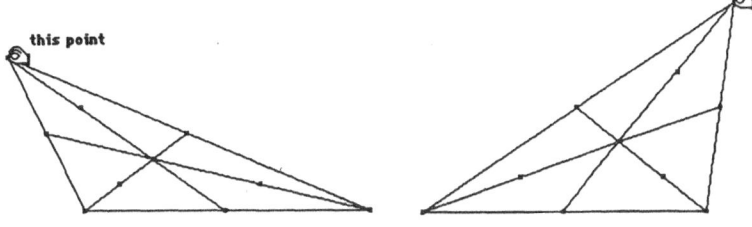

Fig. 3.1–3.2

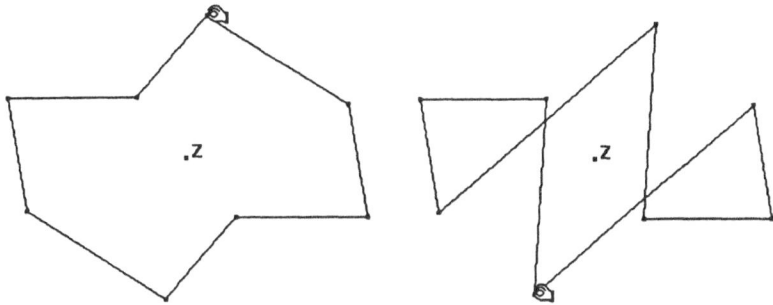

Fig. 4.1–4.2

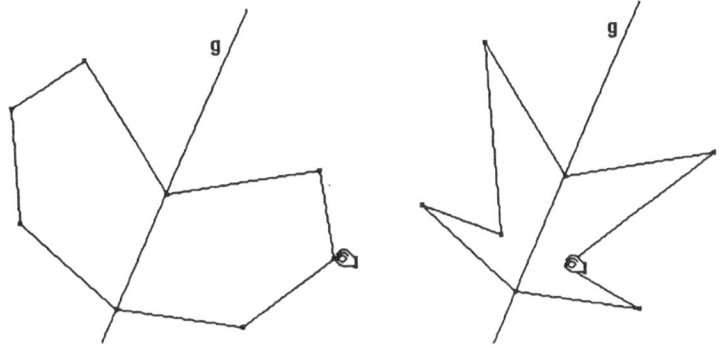

Fig. 5.1–5.2

3.1 Inductive Theorem Finding

Interactive variation of configurations through changing the position of the determining objects (so-called basic objects) in the drag mode. The transition from one state to another is realized continuously (e.g., real time processing) through individual guidance of the dragging cursor. Realization of the operative principle:

The continuous modification of geometric configuration by means of suitable interactive graphics systems facilitates a pure operative orientation of processes regarding theorem finding: Which characteristics of a configuration remain invariant upon the continuous, individually carried out modification process? Thus, theorems from elementary geometry result in statements of invariance upon continuous modification of geometric configurations. The following heuristic strategies can be applied: (direct) analogy, specialization, generalization, borderline cases (see Schumann 1989).

How could the (partial) conjectures regarding invariant properties of configurations as the issue of a mental process by an individual student be checked interactively by the CABRI-Geometer? Version 2.0 of CABRI-Geometer allows the student to select the following properties: incidence, parallelism, orthogonality, isometrism. These properties have still to be extended in respect to the objects being refereed and must be completed by further properties in order to make conjecture-checking more convenient. The program Geometry (Coffey et al. 1986) offers some ideas in this direction. In general this support of theorem finding might diminish the student's interest in theorem proving on a logical level. The great problem of directing the student towards a correct conjecture remains a challenging task.

The preceding possibilities of the variation regarding configurations are explained by the following selected examples.

Example 1: triangle-circumcircle

In Figs. 1.1–1.3, the continuous variation of the triangle with its circumcircle and in the center of the circumcircle M is illustrated. M is the intersection of all mid-perpendiculars. M is located in the interior of the triangle, on the side of the triangle, in the exterior of the triangle, depending whether the triangle is acute-angled, right-angled, or obtuse-angled.

Example 2: relation of the position of median line and mid-perpendicular in a triangle

Figure 6 shows, as in the case of the isolésces triangle, how a bisecting line is located on the mid-perpendicular and, as in the case of the equilateral triangle, how all bisecting lines are situated on the mid-perpendicular.

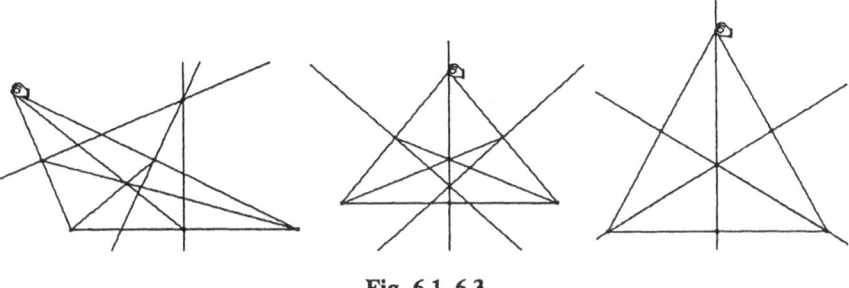

Fig. 6.1–6.3

Example 3: sum of distances for the equilateral triangle

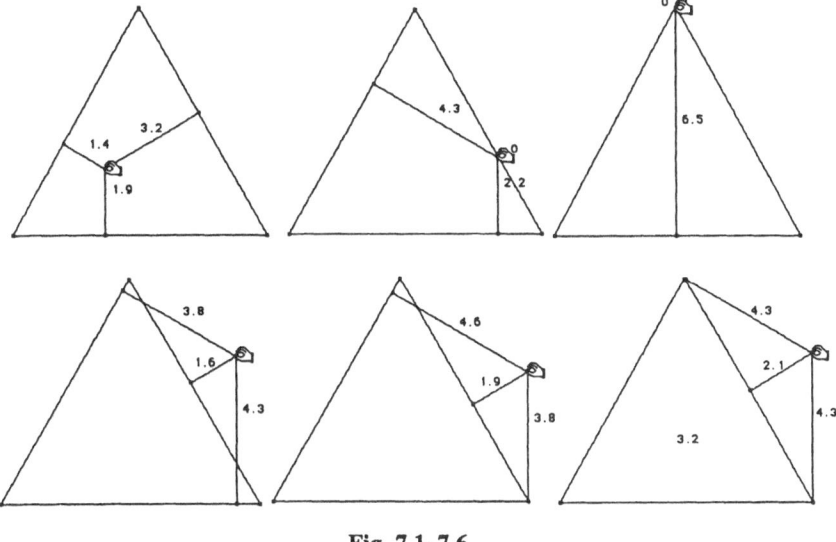

Fig. 7.1–7.6

In Fig. 7.1–7.3, one can see that the sum of distances of a point in the interior or rather on the edge of an equilateral triangle is constant (equal to its altitude). This statement can still be generalized, if the point moves to the exterior of the triangle (see Fig. 7.4–7.5) the perpendicular line segment situated in the exterior has to be taken as negative. The possible position of the point ist limited to the triangle's corners through its perpendiculars (see Fig. 7.6). The geometric location of the border positions forms a regular hexagon.

Example 4: reference quadrangles of bordering squares of a quadrangle

We place squares with their centers on the sides of a quadrangle by means of a suitable macro (Fig. 8.1). We vary the basic quadrangle and observe the quadrangle from the quadrangle centers (Fig. 8.2). If the basic quadrangle becomes similar to a parallelogram, then the reference quadrangle looks like a square (Fig. 8.3); we measure angles and sides again (Fig. 8.4) and correct it as long as the measuring values show a "quadratic" reference quadrangle (Fig. 8.5). The initial quadrangle's shape of a parallelogram can be checked (Fig. 8.6).

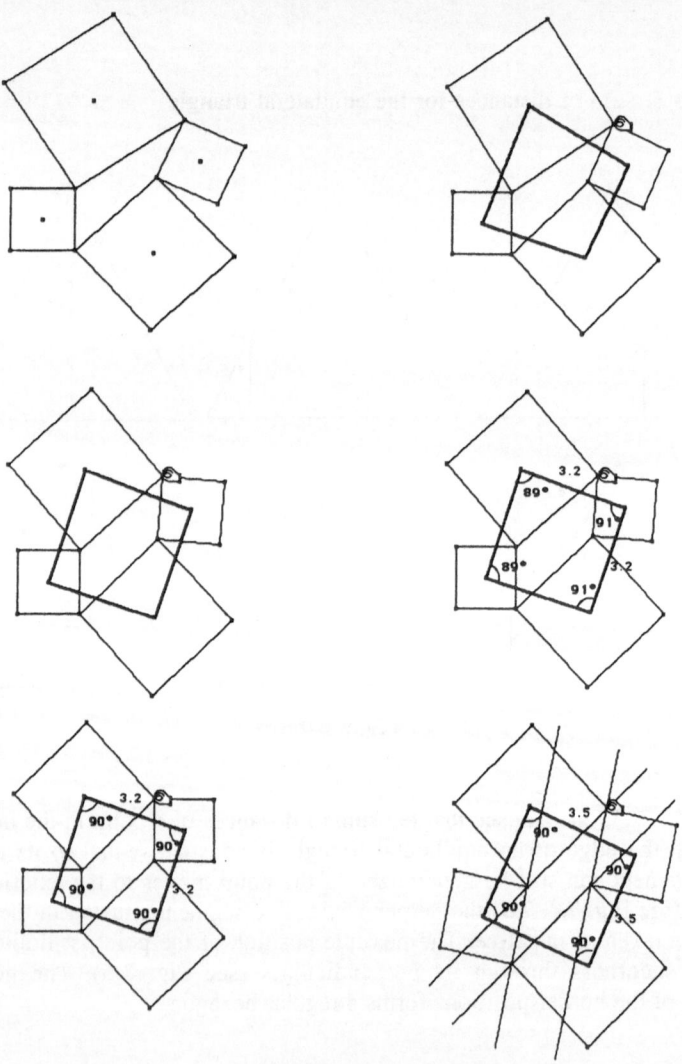

Fig. 8.1–8.6

In the case of the square as a figure for the reference quadrangle, we substantiate our findings by constructing a parallelogram with or without macro – and then by varying the parallelogram (Fig. 9.1–9.2). Even for a degenerate parallelogram the reference quadrangle is a square (see Fig. 9.3). Turning over the parallelogram to the "other side" results in a turning over of the lateral squares to the "interior", the reference quadrangle remains quadratic (see Fig. 9.4). In doing so we have found the theorem of Napoleon–Barlotti for $n = 4$: the connecting n-gon of the centers of the regular n-gons above the sides of an affine-regular n-gon is regular.

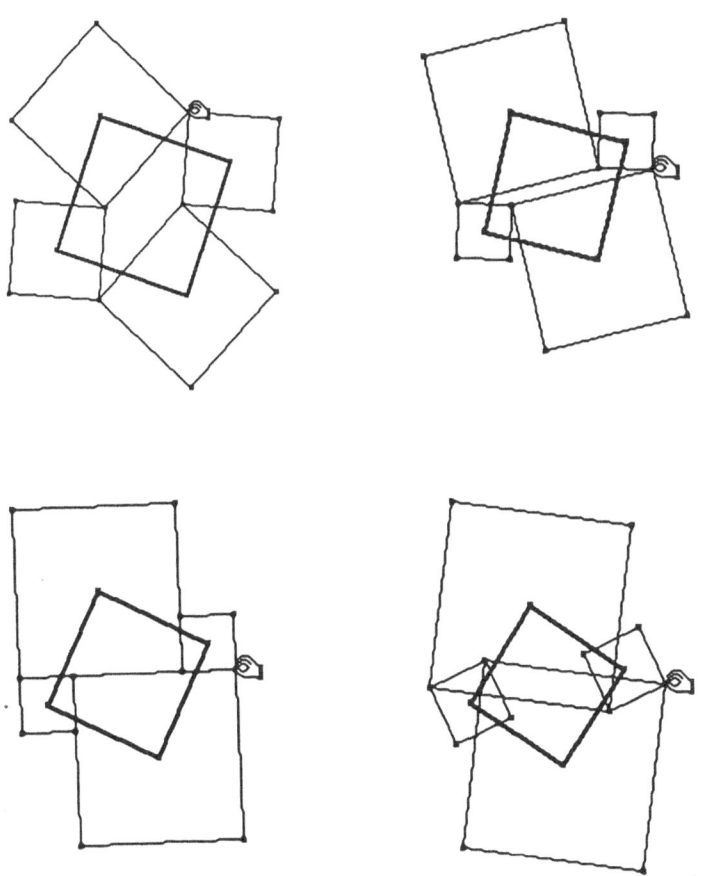

Fig. 9.1–9.4

Now we investigate the diagonals of the reference quadrangle (see Fig. 10.1). When varying the initial quadrangles, the diagonals remain equal in size and orthogonal (see Fig. 10.2–10.3 without measuring values); this is also valid if the initial quadrangle degenerates to a triangle (see Fig. 10.4).

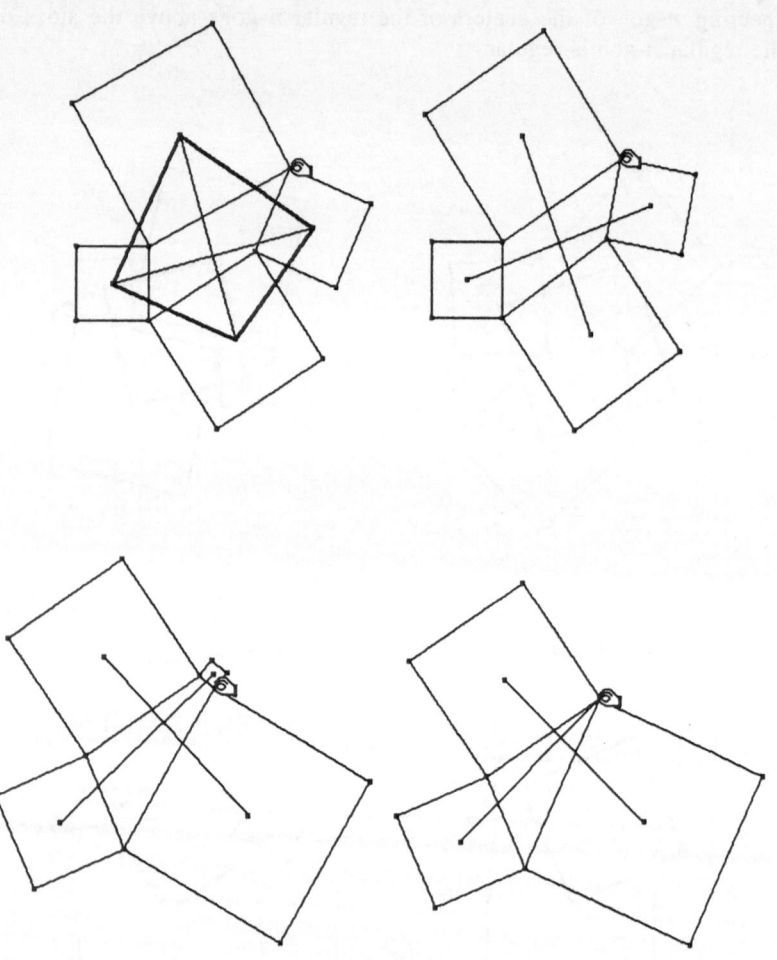

Fig. 10.1–10.4

Alternatively this can be recognized if we are starting from a figure of five equal squares (see Fig. 11.1), separating a corner of the basic square from the construction and dragging it away (see Fig. 11.2). Measurements confirm the invariance of the characteristics regarding the diagonal (see Fig. 11.3–11.4), even for a non-convex quadrangle (see Fig. 11.5).

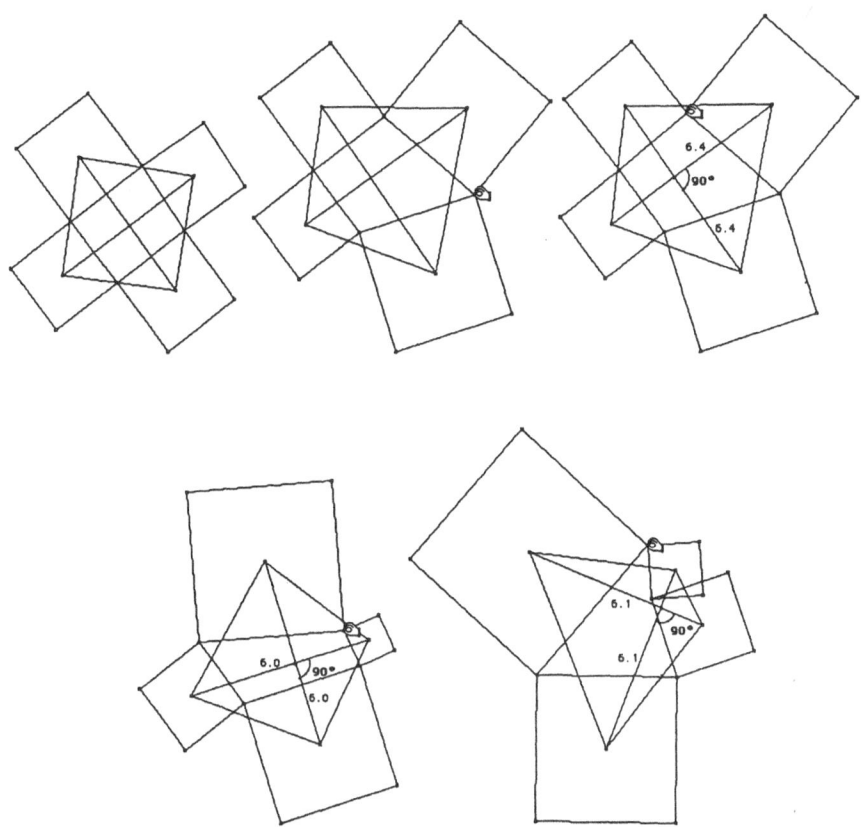

Fig. 11.1–11.5

3.2 Measurement of Line Segments and Angles

Automatic measurement of line segments and angles and dynamic measurement
upon continuous variation of the objects to be measured in the drag mode.

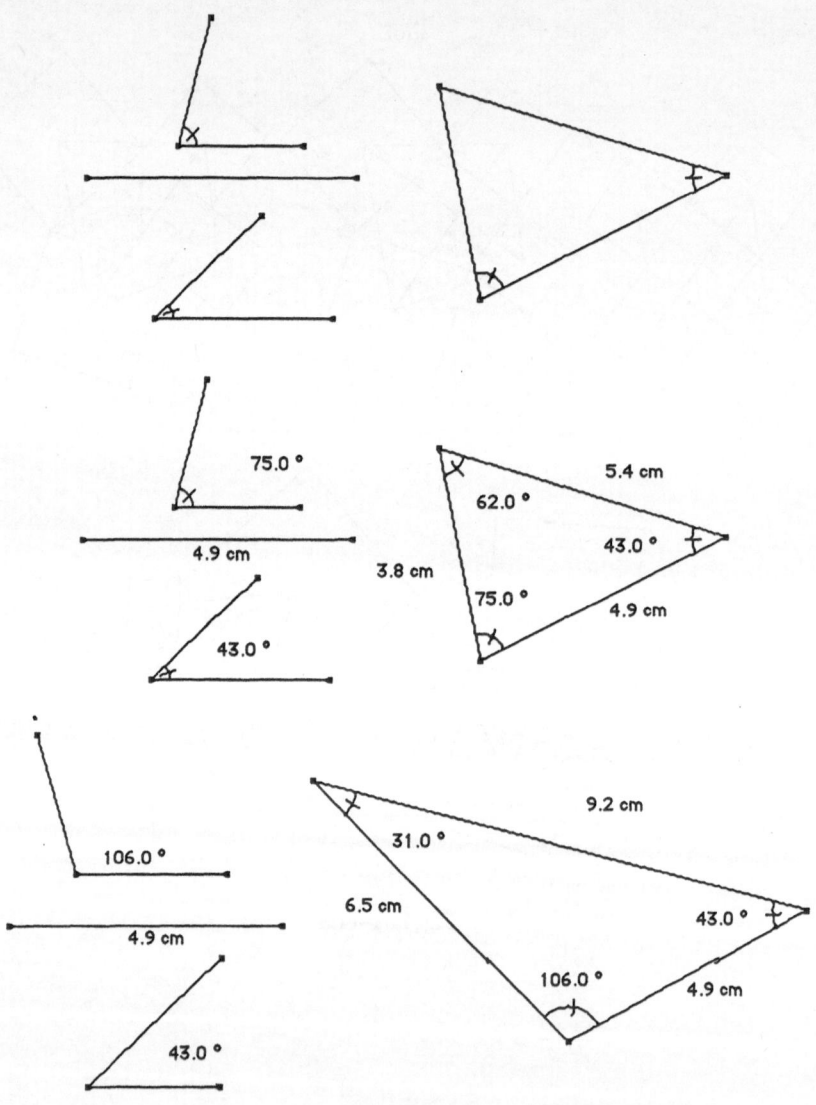

Fig. 12.1–12.3

We illustrate the dynamic measurement of line segments and angles by varying the solving of the triangle problem angle-side-angle. In Fig. 12.1 the result of the construction of a triangle from one side and the adjacent angles are to be seen. We can measure the given objects of the triangle as well as its resulting objects (see Fig. 12.2). Through variation of the externally given objects, the resulting objects on the triangle are varied (see Fig. 12.3). There is a graphic function with input and output parameters, consisting of graphic objects with their measuring values. The interactive variation of input parameters in the drag mode changes the output parameters. Modification of the given side and the given second angle is effected in Fig. 12.4–12.5. If the second angle is made to a 180° complement of the first angle, the triangle degenerates, i.e., its vertex goes into the infinite (see Fig. 12.6).

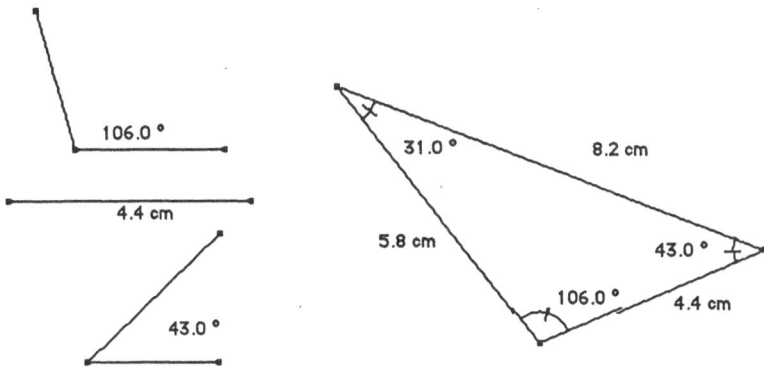

Fig. 12.4

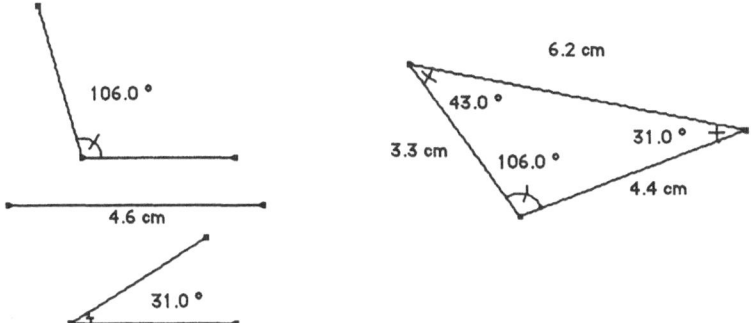

Fig. 12.5

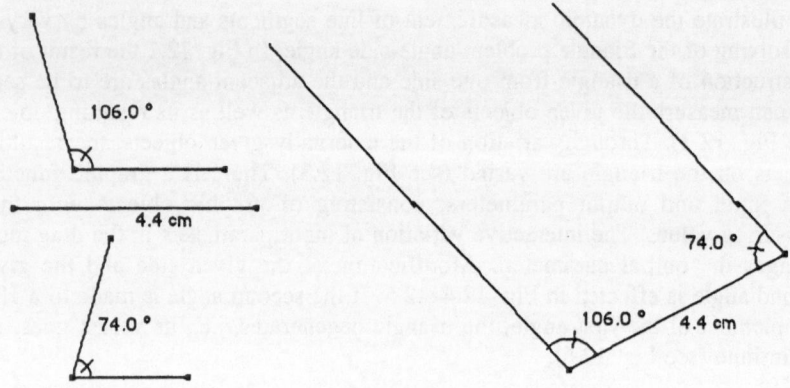

Fig. 12.6

3.3 Generation of Loci

Interactive production of local lines through individual movement of a point on a guiding line, where a point constructionally dependent on it produces the local line point by point. The operational question: which line is described by point Y that is constructionally dependent on X, if X is moved on a guiding line (or even freely) can be investigated.

The interactive production of loci can be efficiently applied: in the heuristic phase of solving construction problems, for the experimental verification of construction results, upon investigations on the position and type of mapped sets, for the construction of algebraic curves of the second order (conic sections) and higher orders, upon investigations on the forms of local lines regarding special points in the triangle.

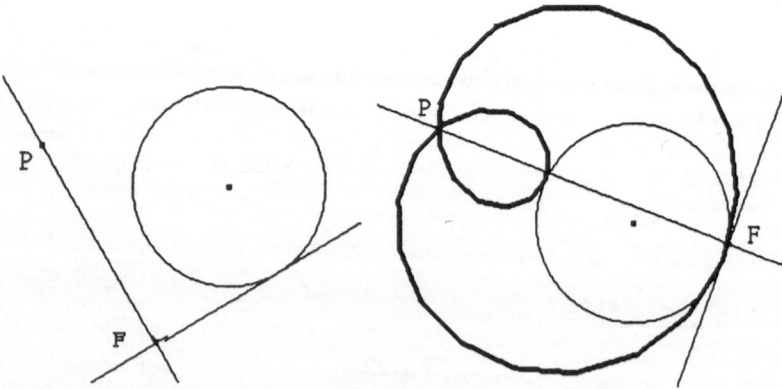

Fig. 13.1–13.2

In order to illustrate the interactive production of local lines, we list three selected examples from the "theory of circles" (Schumann 1990 provides a first didactic systematization).

If the point of contact of a tangent to a circle is moved, the foot of the perpendicular (F) generates a   from the revolution pole on the tangent to a circle (see Fig. 13.1–13.2; curve of the perpendicular foot). Which line describes the point of gravity of a triangle, if a corner of the triangle (here C) is moved on the circumcircle (see Fig. 14.1–14.2, "gravity center circle"). The figure of a circle with an axial affinity is an ellipse; it can be explained at once by the production of a local line (see Fig. 15.1–15.2).

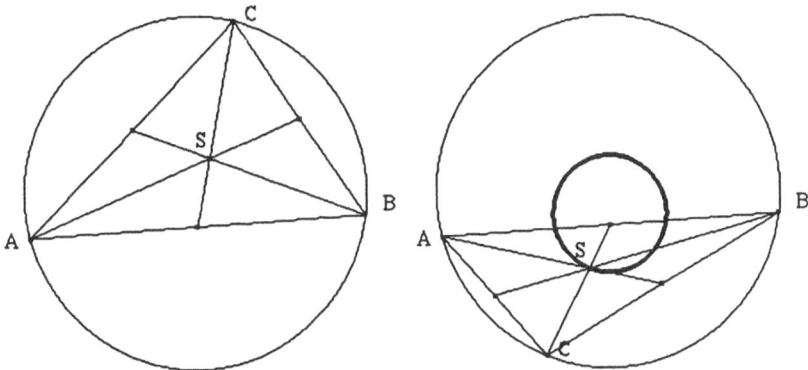

Fig. 14.1–14.2

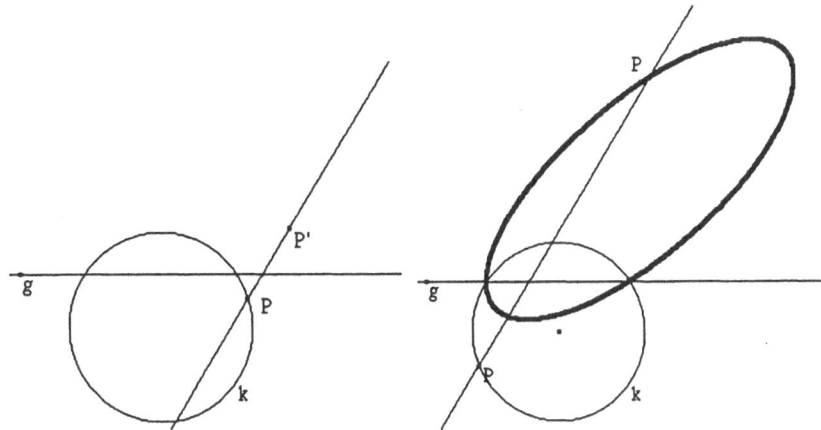

Fig. 15.1–15.2

3.4 Definition and Application of Macro-Constructions

Interactive definition of macro-constructions as graphic functions (through clicking the initial objects and the desired target objects in a drawing; finally input of the individual designation). Interactive application of macro-constructions through clicking of the name of the macro-construction and the initial objects: the target objects as issued for further elaboration. The interactive definition of macro constructions, an essential and versatile tool component, is explained by means of three selected examples.

Example 1: circumcircle of the triangle

Construction of the circumcircle (see Fig. 16.1, facing page). Definition of the initial object triangle; as an alternative, the three corner points or the three (individual) sides can be marked as initial objects. The marking of target objects center and circumcircle is highlighted with dots (Fig. 16.2). Designation of the macro construction (Fig. 16.3). The name of the macro appears at the end of the menu "constructions" and can be selected there (Fig. 16.4). The application of initial objects supplies the desired target objects (Fig. 16.5).

Example 2: product from three reflections

Three straight lines a,b,c (see Fig. 17.1) are given. Reflecting point of P on a is P'; P" is the reflecting point of P' on b and P'" is reflecting point of P" on c. Upon consecutive execution of the reflections on a,b,c; P is reproduced on P'". Initial objects of the reproduced products are the point P and the reflecting straight lines a,b,c; target objects is the point P'" (see Fig. 17.2). The macro "3g-SPIEGELUNG" (reflection) can be applied to each configuration from one point and three straight lines.

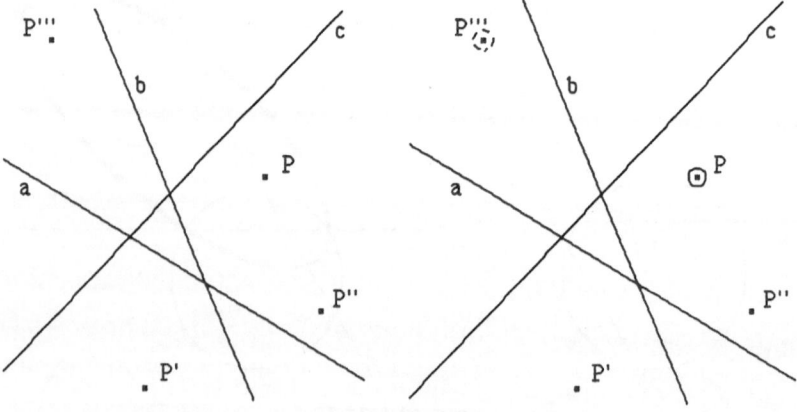

Fig. 17.1–17.2

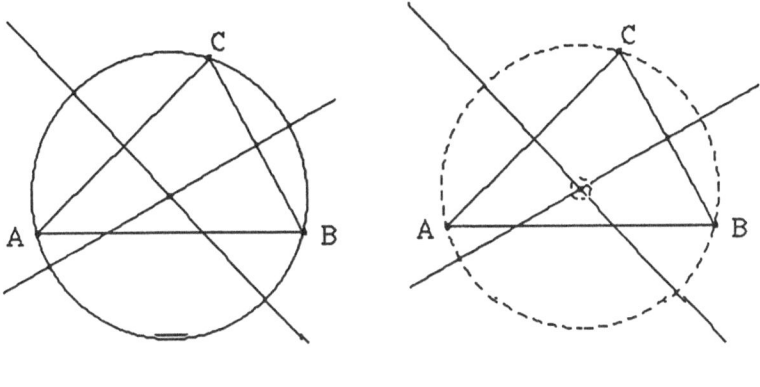

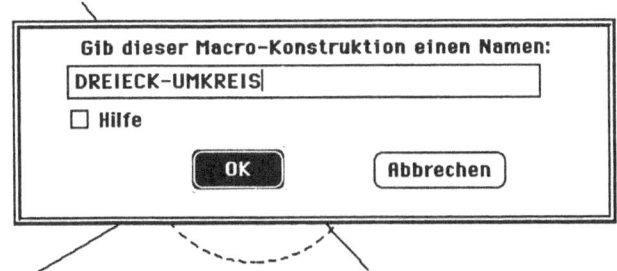

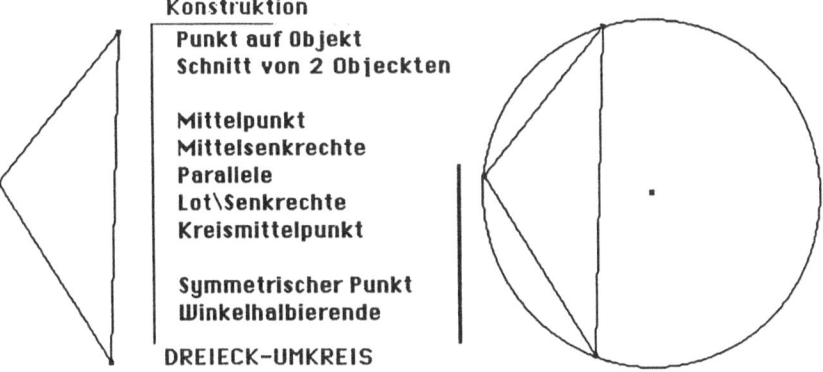

Fig. 16.1–16.5

Example 3: Transformation of a quadrangle into a square equal in area

We start from a quadrangle ABCD (Fig. 18.1) and construct a kite A'BC'D with the same content (Fig. 18.2). For this construction we define a macro; A,B,C,D are initial objects; the sides of the kite are target objects (Fig. 18.3). We conceal all objects as auxiliary lines except for those of the kite (Fig. 18.4). From the kite A'BC'D (Fig. 18.5) we construct a rhombus A'B'C'D' (Fig. 18.6) with the same content and define a corresponding macro (Fig. 18.7). We delete all objects except for the rhombus and construct a square which has the same content as the rhombus (Fig. 18.8) with a corresponding macro (see Fig. 18.9). Now we apply the first macro to a quadrangle (Fig. 18.10) and get a kite of the same content (Fig. 18.11); application of the second macro to the kite (Fig. 18.11) supplies a rhombus of the same content – to obtain a clear overall view, we can conceal the kite (Fig. 18.12); with the third macro (Fig. 18.13) we make a square of the same content from a rhombus and conceal the rhombus (Fig. 18.14). Now, we define the macro, making from each quadrangle a square of the same content, and apply that macro to any quadrangle (Fig. 18.15–18.16). This example demonstrates how a modularized problem solution can be combined to a total solution.

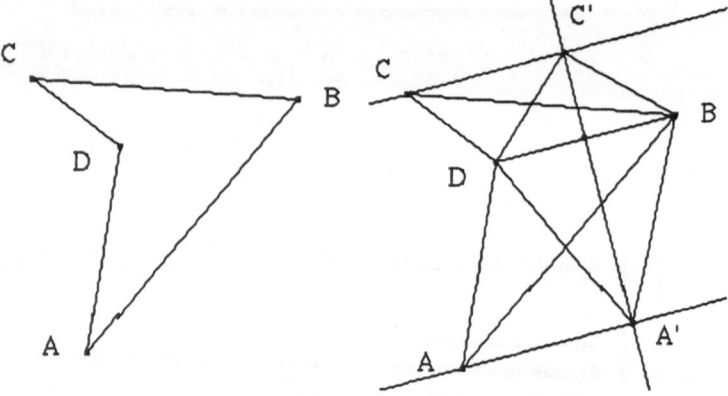

Fig. 18.1–18.2

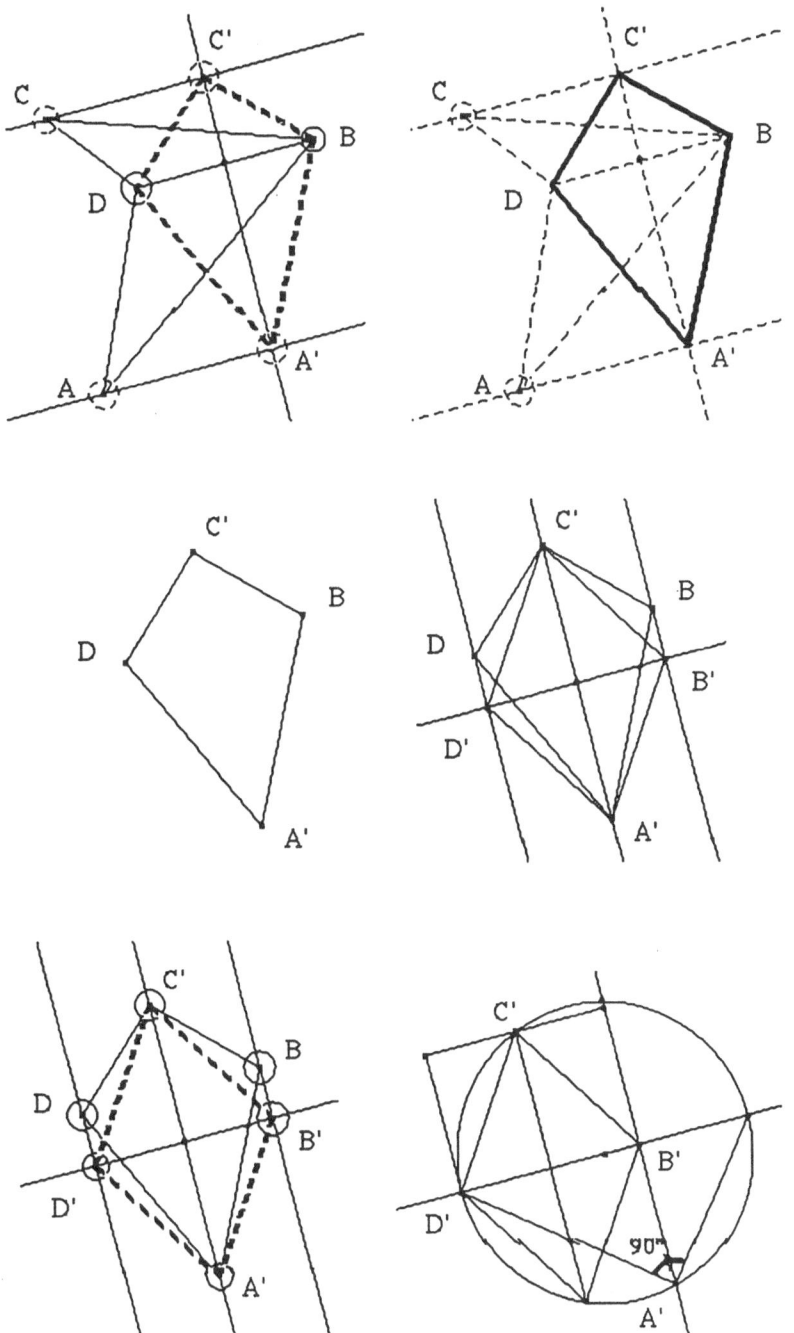

Fig. 18.3–18.8

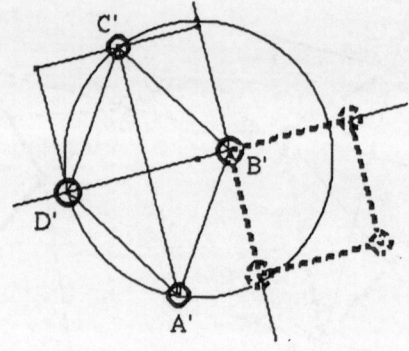

Konstruktion

Punkt auf Objekt
Schnitt von 2 Objeckten

Mittelpunkt
Mittelsenkrechte
Parallele
Lot\Senkrechte
Kreismittelpunkt

Symmetrischer Punkt
Winkelhalbierende

F=VERWDLG. 4-ECK IN DRACHEN.
F=VERWDLG.DRACHEN IN RAUTE
F=VERWDLG.RAUTE IN QUADRAT

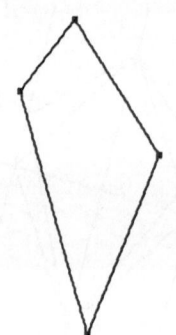

Konstruktion

Punkt auf Objekt
Schnitt von 2 Objeckten

Mittelpunkt
Mittelsenkrechte
Parallele
Lot\Senkrechte
Kreismittelpunkt

Symmetrischer Punkt
Winkelhalbierende

F=VERWDLG. 4-ECK IN DRACHEN
F=VERWDLG.DRACHEN IN RAUTE
F=VERWDLG.RAUTE IN QUADRAT

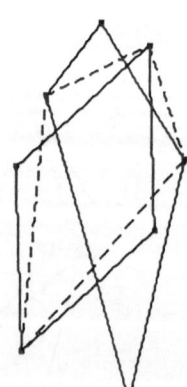

Fig. 18.9–18.12

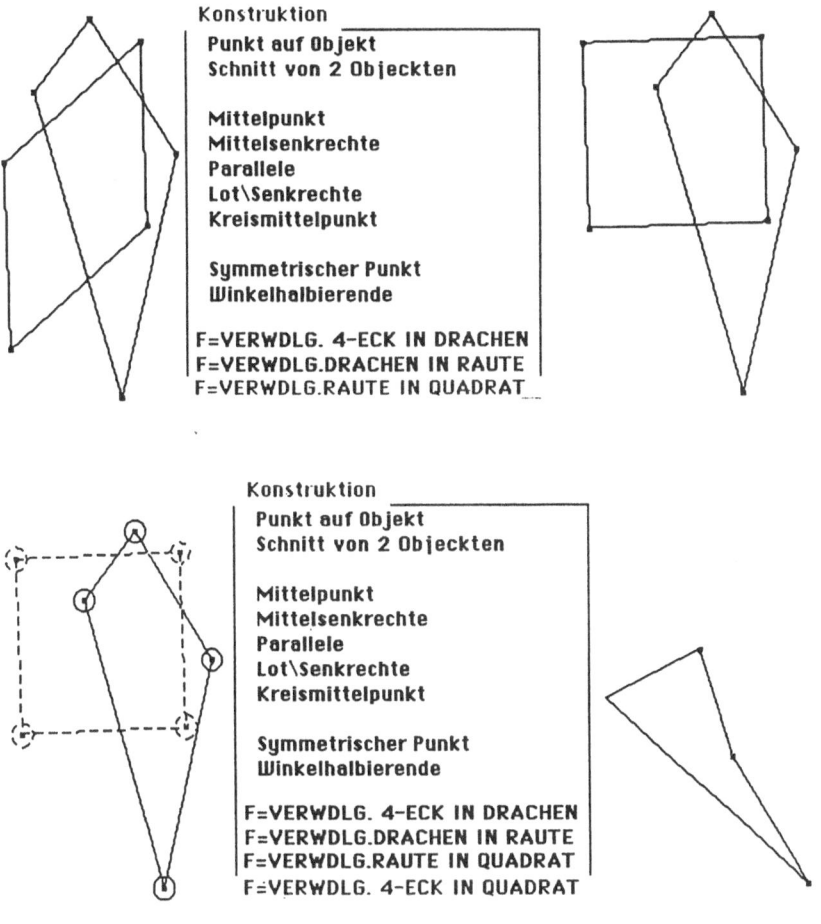

Fig. 18.13–18.16

3.5 The Solution of Planimetric Construction Problems

Especially in the heuristic phase of solving construction problems all possibilities of interactive constructional tools become importing (see Fig. 19). The heuristic phase can be described by the following activities (Holland 1973):
- Drawing a configuration which satisfies the given conditions,
- Varying a configuration and considering (complete) case distinctions
- Establishing first consideration about the determination,
- Supplementing a configuration through drawing of auxiliary lines,
- Reading relations from the configuration,
- Application of heuristic strategies.

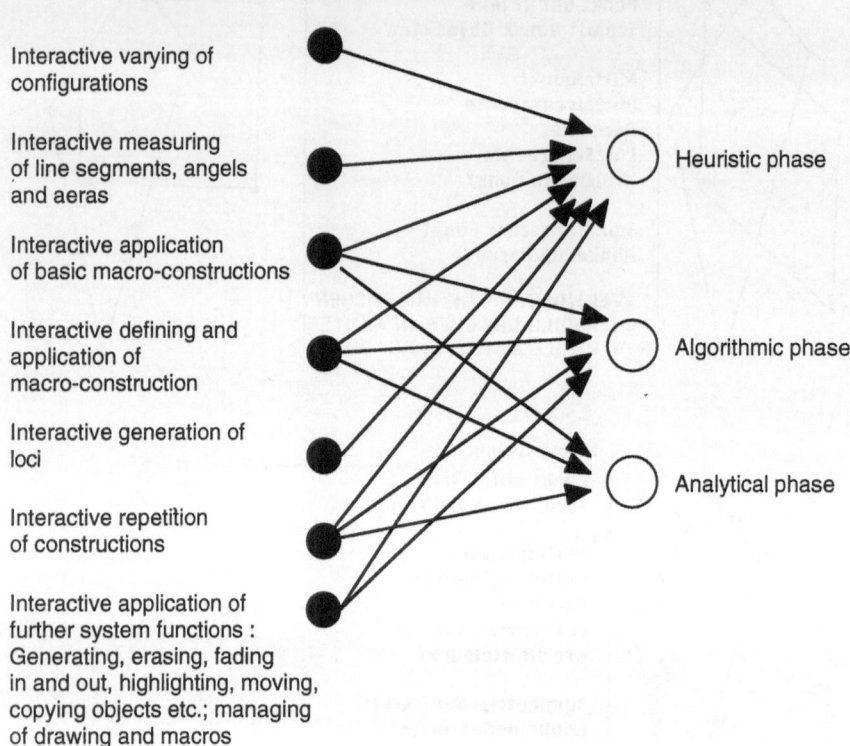

Fig. 19

These activities can be supplemented by virtue of the new tools, e.g., through the production of an experimental solution in the drag mode and/or with interactive production of local lines, which satisfy the given conditions.

The algorithmic phase is characterized by the execution of the found solution way by means of basic constructions, where the solution(s) clearly result from the given objects (basing on geometric statements which are not to explicit). The solution of construction problems can be tested by dragging the basic objects of the construction. Through the definition of a macro construction this phase is to be completed. The repetition mode allows a record of the construction course.

In the heuristic phase as well as in the algorithmic phase, the student is supported by further functions of the system. With regard to the support, the arrows in diagram should be evaluated by means of a representative amount of construction problems.

The analytic phase with regard to the elaboration of a construction problem is not the topic of teaching geometry in the secondary level. It consists of the additional logic explanation with regard to the possibility of pointing out the solution, and of the systematic determination for which synthetic, geometric means are often not more sufficient.

Now we will illustrate the solution procedure supported by interactive construction, especially the heuristic phase by means of an example of an insertion problem.

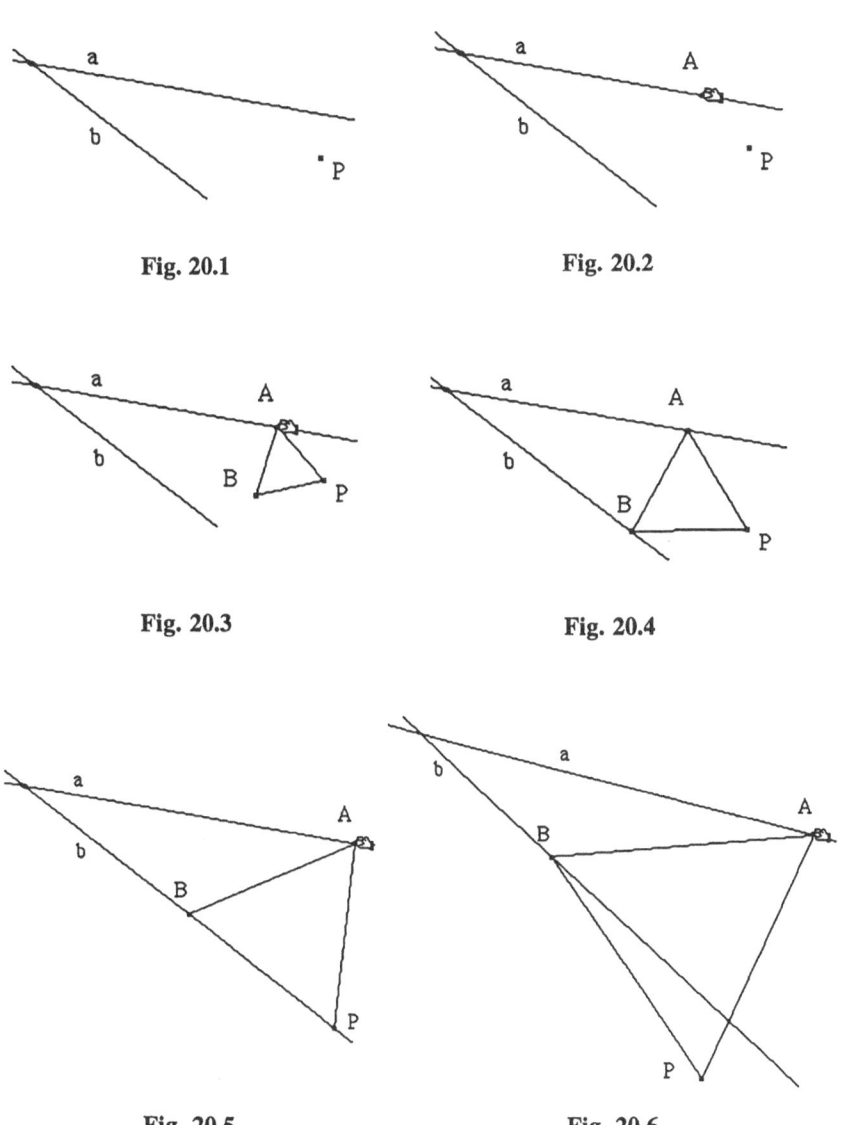

Fig. 20.1   Fig. 20.2

Fig. 20.3   Fig. 20.4

Fig. 20.5   Fig. 20.6

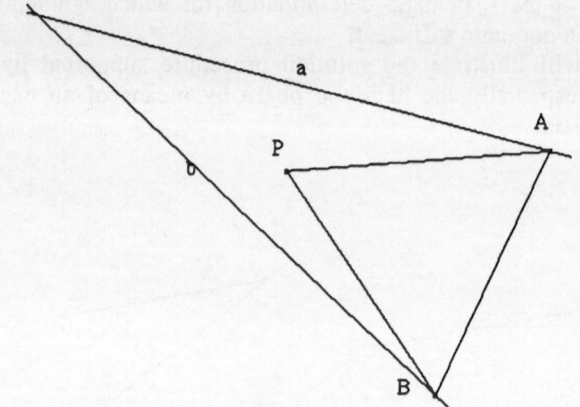

Fig. 20.7

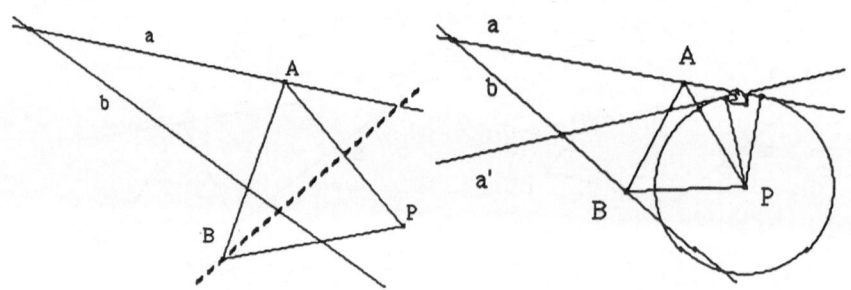

Fig. 20.8   **Fig. 20.9**

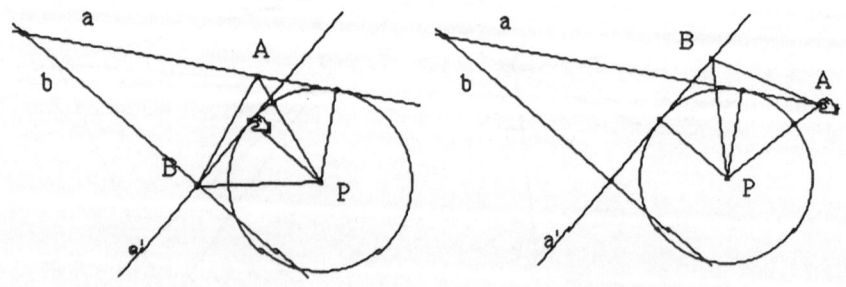

Fig. 20.10   **Fig. 20.11**

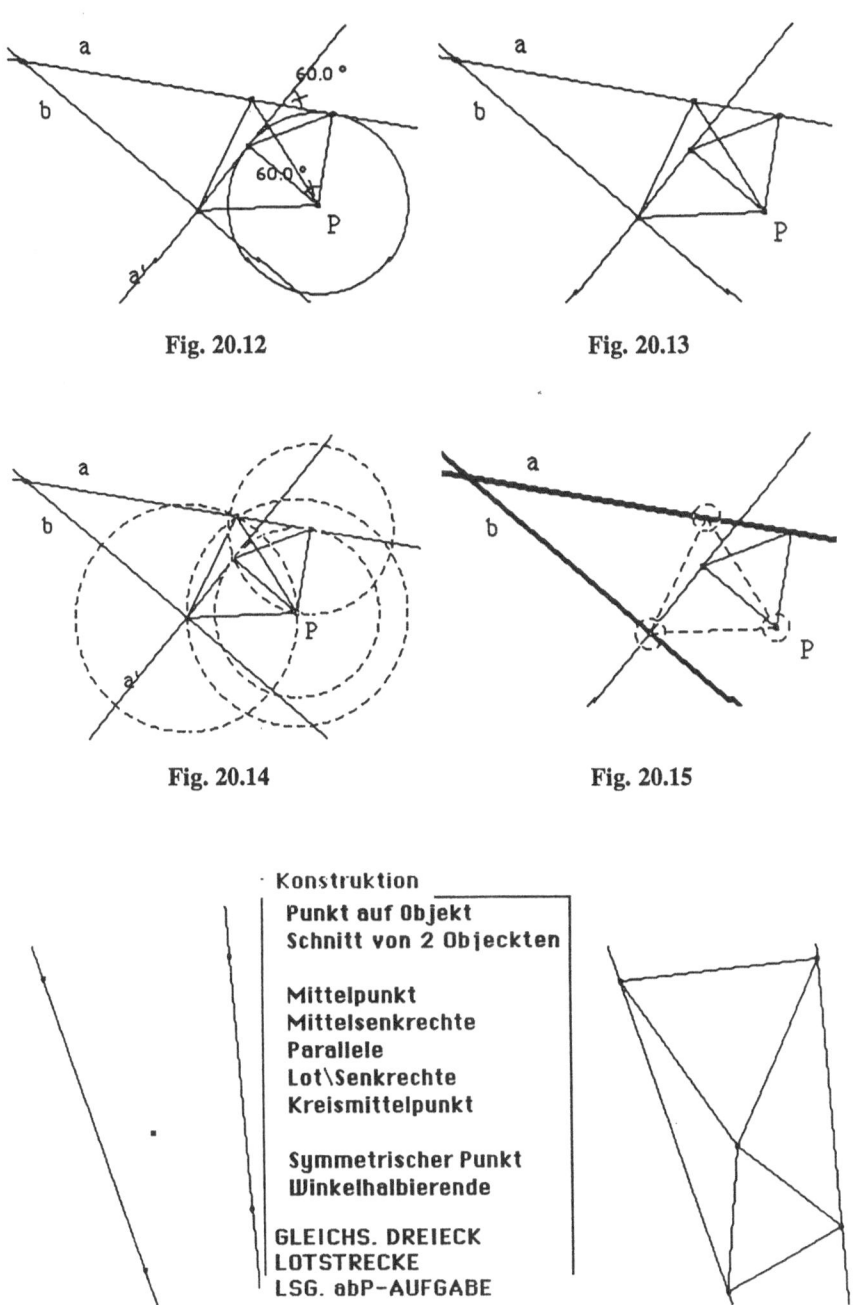

Fig. 20.12

Fig. 20.13

Fig. 20.14

Fig. 20.15

· Konstruktion
   Punkt auf Objekt
   Schnitt von 2 Objieckten

   Mittelpunkt
   Mittelsenkrechte
   Parallele
   Lot\Senkrechte
   Kreismittelpunkt

   Symmetrischer Punkt
   Winkelhalbierende

   GLEICHS. DREIECK
   LOTSTRECKE
   LSG. abP-AUFGABE

Fig. 20.16–20.17

Two straight lines a, b and a point P are given. An equilateral triangle is searched for, which has P as the corner and whose residual corners are on a and b (configuration example in Fig. 20.1). We select a flexible point A on a and construct such a triangle (Fig. 20.2–20.3, heuristic method: leaving out a condition) above PA by means of the macro "GLEICHSEITIGES DREIECK" (equilateral triangle). Variation of the position of A on a produces an experimental solution with B on b (Fig. 20.4). First determination of the solution: Variation of the position of P also allows such a solution (Fig. 20.5–20.6). A further experimental solution is obtained through the application of the macro on AP; P, A, B form a right triple. When moving A on a, leading to an experimental solution (Fig. 20.7), B describes a local line, whose section with b determines the third corner point searched for (Fig. 20.8: local line method, kinematic procedure: the equilateral triangle is revolved and straightened around P). The local line is generated with the option "local line"; it is a straight line. Unfortunately, the method of working back is not completely feasible, since with the local line produced in Fig. 20.8 no further constructions can be carried out. This is a deficiency of the present tool. We try to repair the methodical break caused by the tool, by revolving a around P until the reflected straight line a' goes through B (Fig. 20.9–20.10). For examination, A is moved on a, B has to run on a' (Fig. 20.11), for examination the production of local lines can also be applied (the local line of B has to coincide with a'). The measurement of the angle results in an angle between a and a' of 60°, and a has to be revolved by 60° around P (Fig. 20.12). In doing so, the heuristic phase of the solution finding is completed. The corresponding compass-ruler construction can be shortened with the double application of the macro "EQUILATERAL TRIANGLE" (Fig. 20.13). In the repetition mode the detailed construction with all auxiliary lines can be shown (Fig. 20.14). The algorithmic phase is completed by the definition of a macro with the starting objects a, b, and P (bold in Fig. 20.15) and the target objects (dotted in Fig. 20.15). We designate the macro with "LSG. abP AUFGABE" (problem), apply it to any abP configuration (double) (Fig. 20.16), and vary the position of a,b, and P in order to test the macro (Fig. 20.17 for position variants of P). Now we can study the determination problem more systematically (how many solutions are there dependent on the angle between a and b?)

References

Aebli, H. (1985): Das operative Prinzip (The operational principle). Mathematik lernen, 1, 5-6.

Andelfinger, B. (1988). Geometrie (Geometry). Soest.

Anderson, J.A. et al. (1986). The Geometry Tutor. Journal of Mathematical Behavior, 5, 5-19.

Baulac, Y.; Bellemain, F.; Laborde, J.-M. (1988). Cabri-Géomètre, version 1.0/2.0. - Grenoble 1988/89

Becker, G. (1987). Über den Beitrag des Geometrieunterrichts zum Erwerb heuristischer Strategien (On the contribution of teaching geometry for the acquisition of heuristic strategies). Mathematica didactica, 2, 123-144.

Bender, P. (1987). Anschauliches Beweisen im Geometrieunterricht - unter besonderer Berücksichtigung von (stetigen) Bewegungen bzw. Verformungen. (Illustrative proving in teaching geometry – under special consideration of (continuous) movement or deformations respectively). Preprint no. 1, Kassel.

Chazan, D.; Houde, R. (1989). How to use conjecturing and microcomputers to teach geometry. Reston.

Coffey, M. et al. (1986). Geometry. San Rafael.

Holland, G. (1973). Die Bedeutung von Konstruktionsaufgaben für den Geometrieunterricht (The significance of construction problems for teaching geometry). Beitrage zum Mathematikunterricht, 11-24.

Holland, G. (1989). GEOCON, Eine lernfähige Lernumgebung fiir geometrische Konstruktionen (GEOCON, a learning environment conducive to learning for geometric constructions). Beiträge zum Mathematikunterricht 203-206.

Moore, O.K.; Anderson, A.R. (1969). Some principles for the design of clarifying educational environments. In: Goslin, A. (ed.): Handbook of Socialization Theory and Research. Chicago.

Schumann, H. (1988). Der Computer als Werkzeug zum Konstruieren im Geometrieunterricht (The computer as tool for construction in geometric education). Zentralblatt fur Didaktik der Mathematik, 6, 248-263.

Schumann, H. (1990). Interaktives Erzeugen von Ortslinien – ein Beitrag zum computerunterstützten Geometrieunterricht (Interactive production of local lines – a contribution to computer support geometric education). Mathematik lernen, 1.

Schumann, H. (1989). Neue Möglichkeiten des Geometrielernens durch interaktives Konstruieren (New possibilities of learning geometry through interactive construction). In: Research Report 74, University of Helsinki "Conference on the Teaching of Geometry" Helsinki 1.4.8, pp. 265-272.

Schumann, H. (1989). Satzfindung durch kontinuierliches Variieren geometrischer Konfigurationen mit dem Computer als interaktivem Werkzeug (Theorem finding through continuous variation of geometric configurations with the computer interactive tool). Der Mathematikunterricht, 4, 22-37.

Schumann, H. (1989). The computer as a tool for geometric constructions. Micromath, 3, 53-56.

Schwartz, J.L., Yerushalmy, M. (1985): The Geometric Supposer: Triangles. Pleasantville.

Winter, H. (1981). Entdeckendes Lernen im Mathematikunterricht (Discovering learning in mathematical education), Braunschweig.

Wittmann, E. (1981). Grundfragen des Mathematikunterrichts (Basic questions of teaching mathematics). Braunschweig.

Yerushalmy, M. (1986). Induction and generalization: an experiment in teaching and learning high school geometry. Ann Arbor, MI.

Socratic Tutoring with Software:
An Example and a Prospectus

Judah L. Schwartz

Professor of Engineering Science and Education
Massachusetts Institute of Technology, Cambridge, MA, USA
&
Co-Director, Educational Technology Center
Graduate School of Education, Harvard University, Cambridge, MA 02138, USA
E-mail: judah@hugse1.harvard.edu

Abstract. This paper describes a procedure for algorithmically generating a connected semantic graph that captures the structure of any algebraic modeling problem as it is being solved. The generated graph then allows a microcomputer-based tutor to guide a student in traversing any logically permissible path through the space of semantic relations of the problem situation. The generalizability of this approach to ICAI in geometry and other mathematical domains is discussed.

Introduction

There is by now an extensive literature on the use of computer based intelligent tutors (see for example Sleeman & Brown 1982, White & Frederikson, 1986). For the most part, the research in this area can be characterized as having the following shared set of assumptions.
  – The computer based tutor ought to have built into it some model of the problem domain.
  – Often the model of the domain includes a model of 'expert' behavior within the domain.
  – The computer based tutor ought to have built into it some sort of a model of the student.
  – The tutor ought to use these two models along with the student's input to decide, at each step of its interaction with the student, what next to 'do' or 'say'.
Over the past two decades or so, different researchers have implemented a variety of Intelligent Computer Assisted Instruction (ICAI) systems in widely varying subject matter domains. Despite the variety of disciplines and topics addressed by these systems and the varied audiences for which they were intended, it seems to me that the vast majority of these systems share the assumptions listed above.

This is a paper about a different kind of computer-based tutor, one that derives from a different set of assumptions. The impetus for the development of this tutor is a desire to build software that can tutor students intelligently in some non-trivial domain but that is consistent with the author's educational ideology. The essence of this ideology is that human behavior is sufficiently complex, and our understanding of it sufficiently fragile, that it is foolish to imagine that, in the near term at least, we will be able to build intelligent software that depends for its operation on being able to infer the intention of the user. Given this ideological stance, and the desire to build intelligent tutorial software, one is led to consider a somewhat different approach to the problem of ICAI than is traditionally taken.

There is a second impetus for the development of the tutor that this paper reports on. One of the traditional assumptions of ICAI researchers has been the need for the computer-based tutor to have a model of the problem domain. While desirable, it seems to me to be a formidable task, given the vast array of problem domains that we as humans come to learn something about. Wouldn't it be nice to devise a tool that could help people trying to solve problems in quantum mechanics, economics, chemical kinetics, high school algebra and fourth grade word problems without having to build expertise in each of these areas into the tool?

In this paper we will report briefly on a modest computer-based tutor for people using an algebraic modeling environment called the ALGEBRAIC PROPOSER (Schwartz, 1987). We will then attempt to analyze how the principles employed in the design and implementation of that environment can be used in geometry and other domains.

The following constraints on the design of the tutor were adopted:

– The tutor should make no inferences about the tutee's intentions; it follows therefore, that the tutor must be able to function without a theory of the learner.

– The tutor should be generative in the sense of being able to guide students through problem trajectories that are different from, but logically equivalent to, those formulated, and/or expected, by the problem poser.

– The problem domain of the tutor should encompass any quantitative problem, in any subject area, that can be formulated algebraically.

A Problem

To make the discussion concrete, at least at the outset, we will focus our attention on a particular problem which we shall analyze in several different ways. The analysis of the problem will also serve to introduce the ALGEBRAIC PROPOSER environment. Here is the problem.

Olga and Teresa set out to mow a lawn together. When Olga mows the lawn alone it takes her 2 hours to do so. When Olga and Teresa mow the lawn together, it them only 3/4 of an hour. How long would it take Teresa, working alone, to mow the lawn?

This problem is appropriate to, but is normally considered difficult for, high school students studying algebra. In order to get started, let us list the known numerical quantities in the problem.

	How many	What		Notes
A	1	lawn		the size of the job to be done
B	2	hr		time for Olga to mow the lawn
C	0.75	hr		mowing time for Olga and Teresa

In addition, let us put on our list the unknown quantity, T hr, i.e., the time it takes Teresa, working alone, to mow the lawn.

D	T	hr		time for Teresa to mow lawn

The key to solving the problem is the recognition that the quantities to be combined by virtue of Olga and Teresa working together are either the mowing rates (i.e. the lawns/hour) of Olga and Teresa or the fraction of lawn (i.e. lawns) that each of them mows. One does not solve the problem by combining the time Olga mows with the time Teresa mows.

What are the mowing rates of Olga and Teresa? Let us calculate them and add them to our list.

E	0.5	lawn/hr	A/B	mowing rate of Olga
F	1/T	lawn/hr	A/D	mowing rate of Teresa

The combined mowing rate of 1 and 2 working together is

G	0.5 + 1/T	lawn/hr	E+F	combined mowing rate for the two

At this point there are three different but equivalent ways of proceeding. Each way entails making a statement about some aspect of the problem being modeled. We can characterize these different ways as:

i) constraining 'hours'
ii) constraining 'lawns/hour'
iii) constraining 'lawns'

We will analyze the first of these possibilities and defer consideration of the others. If we divide the mowing job to be done, i.e., 1 lawn, by the combined mowing rate of Olga and Teresa, i.e., 0.5 + 1/T hr, we get an expression for the time it takes them to mow the lawn together.

H	1/(0.5 + 1/T)	hr	A/G	mowing time for Olga and Teresa

We can require that this quantity be equal to the know time that it takes Olga and Teresa to mow the lawn, i.e., 0.75 hr. If we do so we obtain the equation

$$1 / (0.5 + 1/T) = 0.75 \text{ hr}$$

Our screen now looks like this:

	How many	What		Notes
A	1	lawn		the size of the job to be done
B	2	hr		time for Olga to mow the lawn
C	0.75	hr		mowing time for Olga and Teresa
D	T	hr		time for Teresa to mow lawn
E	0.5	lawn/hr	A/B	mowing rate of Olga
F	1/T	lawn/hr	A/D	mowing rate of Teresa
G	0.5 + 1/T	lawn/hr	E+F	combined mowing rate for the two
H	1/(0.5 + 1/T)	hr	A/G	mowing time for Olga and Teresa

Comparing functions

$$A \quad 1 / (0.5 + 1/T) = 0.75 \text{ hr}$$

The reader will note that the ALGEBRAIC PROPOSER keeps track of, and distinguishes between, user-entered quantities and user-computed quantities.

Had our strategy been to constrain the mowing rate [in units of lawns/hr] rather than the time, then we could calculate an expression for the combined mowing rate at which the job is done, i.e., 1 lawn /0.75 hr, or:

$$1.33333 \text{ lawn/hr.}$$

We can require that this quantity be equal to the combined rate of Olga and Teresa mowing together, i.e., (0.5 / 1/T) lawn/hr. Doing so we obtain we obtain the equation,

$$0.5 + 1/T = 1.33333 \text{ lawn/hr}$$

Had we chosen to constrain the amount of work done to be equal to the sum of the amounts done by Olga and Teresa then we can multiply the combined mowing rate of Olga and Teresa, i.e., 0.5 + 1/T lawns/hr by the time it takes them working together to mow the lawn to get

$$0.75 * (0.5 + 1/T) \text{ lawn}$$

We can then require that this quantity be equal to the mowing job that needs to be done, i.e., 1 lawn. Doing so yields the equation

$$0.75 * (0.5 + 1/T) = 1 \text{ lawn}$$

Our screen now looks like this:

	How many	What		Notes
A	1	lawn		the size of the job to be done
B	2	hr		time for Olga to mow the lawn
C	0.75	hr		mowing time for Olga and Teresa
D	T	hr		time for Teresa to mow lawn
E	0.5	lawn/hr	A/B	mowing rate of Olga
F	1/T	lawn/hr	A/D	mowing rate of Teresa
G	0.5 + 1/T	lawn/hr	E+F	combined mowing rate for the two
H	1/(0.5 + 1/T)	hr	A/G	mowing time for Olga and Teresa
I	1.33333	lawn/hr	A/C	combined mowing rate for the two
J	0.75 * (0.5 + 1/T)	lawn	C*G	sum of amounts of lawn mowed by Olga and Teresa.

Comparing functions

$$\text{A} \qquad 1 / (0.5 + 1/T) = 0.75 \ \ \text{hr}$$
$$\text{B} \qquad 0.5 + 1/T = 1.33333 \ \ \text{lawn/hr}$$
$$\text{C} \qquad 0.75 * (0.5 + 1/T) = 1 \ \ \text{lawn}$$

Thus we see three quite different analyses, each of which focuses on a different attribute of the referent situation. Each analysis leads to a different equation. Obviously all of the equations are members of the same equivalence class of equations and derivable from one another algebraically.

It is entirely likely that different people will find it congenial to focus their attention on different attributes of this situation. As a consequence, they are likely to compute different quantities entailed by the givens and write different equations.

If we were simply designing a tutor to work with this particular problem, or for that matter, a more general tutor that works with the class of problems to which this problem belongs, it would presumably not be difficult to address the problem of dealing with the fact that different individuals will take different pathways through the problem and find different features of it relatively more salient. To do this however, would require some sort of modeling of the problem domain, something that we explicitly wish to avoid.

Graphing the Semantic Space of the Problem

In the problem analyzed above, the quantities A, B and C are quantities that are referred to in the problem statement both by name and by value. The quantity D is a quantity referred to in the problem statement by name but not by value. The quantities E, F, G, H, I and J are computed in terms of the user-entered quantities A through D. The software uses this information to compute the following directed graph of the structure of the relationships among the quantities, both user entered and computed.

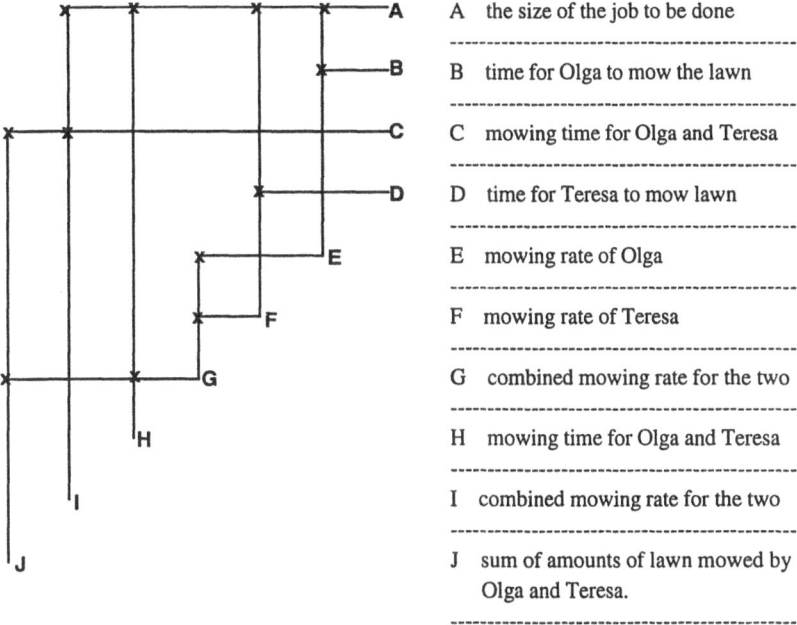

A   the size of the job to be done

B   time for Olga to mow the lawn

C   mowing time for Olga and Teresa

D   time for Teresa to mow lawn

E   mowing rate of Olga

F   mowing rate of Teresa

G   combined mowing rate for the two

H   mowing time for Olga and Teresa

I   combined mowing rate for the two

J   sum of amounts of lawn mowed by
     Olga and Teresa.

Comparing functions

$$
\begin{array}{ll}
A & H = C \\
B & G = I \\
C & J = A
\end{array}
$$

The three equivalent equations that were derived above involve comparing, via the equality symbol, the functions H and C each of which is a time, or comparing the functions G and I each which is a mowing rate, or comparing the functions J and A each of which is an amount of work.

The Tutor and the Semantic Graph

Clearly, there are many possible ways for a student to traverse this semantic graph starting from the user-entered quantities that are referred to in the problem statement and arriving at a solution to the problem.

> *N.B. By solution, I mean here a solution to the modeling problem, and not to the subsequent symbolic manipulation problem that results in a numerical value for the time.*

How can we guide students along an acceptable path through the semantic graph? The key to a resolution of the problem lies in the notes that appear on the right hand side of the quantities in the ALGEBRAIC PROPOSER, each of which constitutes a verbal representation of either an entered or a computed quantity.

Suppose that I, as a teacher had entered the problem under discussion into the ALGEBRAIC PROPOSER, complete with notes on the quantities, as shown above. Let us further assume, that I have stored the problem in a file called WORK. A student who asks for help with this problem activates the built in tutor, which reads the file and computes its semantic graph. The tutor then offers the students a display of the semantic graph and the accompanying notes. In addition, the tutor isolates the entered quantities known by both name and value and prompts the student to

enter the KNOWN quantities
A   the size of the job to be done
B   time for Olga to mow lawn
C   mowing time for Olga and Teresa

The student responds by choosing A, B or C. Let us say he chooses B, thereby indicating that he wishes to enter the quantity "time for Olga to mow lawn" If the student enters the quantity 2 hr, the entry is accepted and placed in the appropriate places in the *How many* and *What* columns. Entering a different quantity at this point, even if it is one of the other given quantities of the problem will cause the tutor to reject the response.

After the student has entered all of the known quantities in the problem, the tutor prompts the student to enter the entered quantities that are known by name but not by value, i.e., the variables, in much the same way.

After the known and variable quantities have been entered, the tutor then offers the message

you can now write an expression for
E   mowing rate of Olga
F   mowing rate of Teresa
I   combined mowing rate for the two

Note that these are the *only* quantities that are computable with the information the student has available at the moment, and thus these are the only quantities on the prompt list at the moment. If the student chooses F, for example, the tutor will ask what two quantities and what operation are necessary to compute the quantity F. If the student enters A/D the response is accepted, the quantity 1/T hr is entered in the appropriate places in the *How many* and *What* columns, and the entry "F mowing rate of Teresa" is removed from the prompt list. It should be noted that after the student states his intention to write an expression for the "mowing rate of Teresa" any response other than A/D will be rejected by the tutor.

Assume the student responds correctly for the quantity F and then responds correctly for the quantity E. As soon as the quantities E and F are correctly entered, it becomes logically possible to compute the quantity G. At that time the quantity G is immediately added to the prompt list.

Because the full graph and all the accompanying notes are available to the student throughout the session, the student can reason by backward chaining from the goal state(s) that can be read from the graph. In addition the student can traverse the graph of the problem in a 'forward' direction in any way that is logically acceptable. The student's path through the problem space need not coincide with the path taken by the instructor when the problem was entered. Indeed, it is possible that the student might take a path that the instructor did not anticipate.

The tutor clearly does not have a model of the student, nor does it have any sort of a priori semantic model of the problem. It makes no inference about the student's computational intentions. By simply presenting the range of logically acceptable alternatives at each step of the problem solving process, it can suggest possible next steps to an otherwise floundering student.

In summary, the essence of this tutoring style is to ask the student to enter the known quantities, both magnitudes and referents, of the problem. The student is prompted by short verbal descriptions of the known quantities. After that, the student is prompted to enter the unknown quantities of the problem. Here the magnitudes are entered symbolically while the referents are entered explicitly.

The tutor then offers the student a list of the then logically calculable quantities. The student chooses which of these quantities he or she wishes to calculate. If the student correctly indicates how to calculate the quantity he or she has chosen, the quantity is added to the list of "computed" quantities and removed from the prompt list. The process continues until all the quantities necessary for the construction of the requisite equations or inequalities of the problem are on hand.

The tutor then indicates which referents can be constrained by the formulation of an equation or inequality. Again, the student chooses a quantity to constrain. If the subsequent equation (or inequality) constraining the quantity coincides with the one written for that quantity by the preparer of the problem, it is removed from the prompt list. Again, the process repeats until all the necessary equations and/or inequalities have been written.

From Particularity to Generality

The potential power of this tutoring procedure based on the semantic graph of the situation lies in its ability to cause people to think about problem structure. To the extent that it succeeds in doing so, it frees students of the narrowness of the particular problem on which they are working and allows them to think more generally about the class of problems to which the particular problem belongs. Here is an example.

A commuter is running to catch a train at the station. He is running at a speed of 8 m/s and is 30 m from the train at the instant the train begins to accelerate from rest with a constant acceleration of 1 m/s*s. He continues to run at 8 m/s toward the train while the train accelerates. What is the minimum time in which the commuter can catch the train? How long after the train starts accelerating is it no longer possible for the commuter to catch the train?

A person using the ALGEBRAIC PROPOSER who seeks to model this situation might solve the problem as follows:

	How many	What		Notes
A	8	m/s		speed of running commuter
B	30	m		initial train-commuter separation
C	1	m/s*s		acceleration of train
D	T	s		time variable
E	8T	m	A*D	distance of commuter from start at time T
F	1T	m/s	C*D	instantaneous speed of train at time T
G	1/2			constant needed to compute average speed
H	1/2(1T)	m/s	G*F	average speed of train during time T
I	1/2(1T)T	m	H*D	distance train travels in time T
J	1/2(1T)T+30	m	I+B	distance of train from start at time T

If the commuter is to be able to board the train then the quantities

    J   distance of train from start at time T

and

    E   distance of commuter from start at time T

must be constrained to have some relationship to one another. If we assume that the position of the train is measured from the rear of the train then we require that the quantity E be greater than the quantity J if the commuter is to catch the train. This leads us to the inequality

$$8T > 1/2(1T)T + 30 \text{ m}$$

The values of T that satisfy this inequality bound the values of T for which it is possible for the commuter to catch the train.

Here is the semantic graph that the ALGEBRAIC PROPOSER generated:

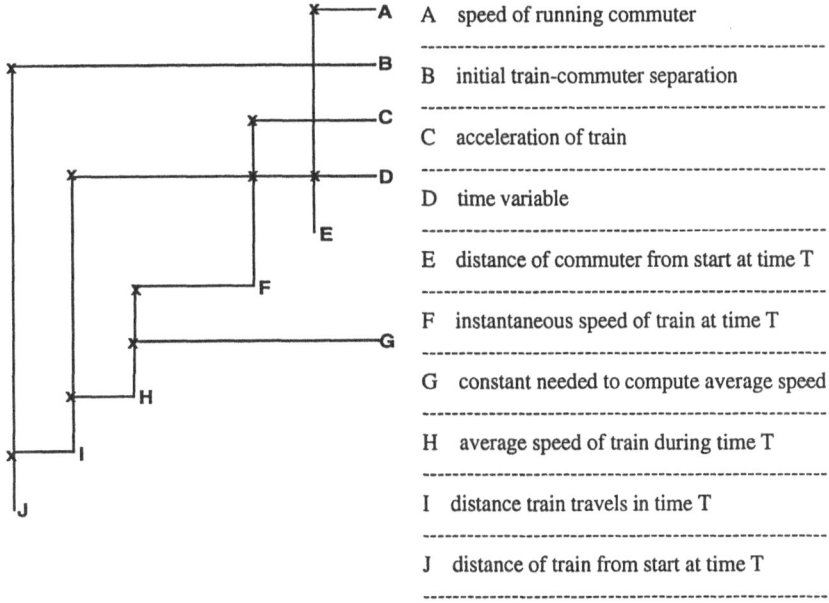

A   speed of running commuter

B   initial train-commuter separation

C   acceleration of train

D   time variable

E   distance of commuter from start at time T

F   instantaneous speed of train at time T

G   constant needed to compute average speed

H   average speed of train during time T

I   distance train travels in time T

J   distance of train from start at time T

Since a primary strategy for solving problems of this sort depends on writing two different functions that have referents that are known to be related to one another, E and J in this instance, it is possible to see immediately in the entailment map that the equation or inequality that 'solves the problem' is one that constrains the relationship between these two quantities.

Now suppose that we consider a problem that most students would regard as quite different.

> *A commuter is running to catch a train at the station. He is running at a speed of V m/s and is 30 m from the train at the instant the train begins to accelerate from rest with a constant acceleration of 1 m/s*s. He continues to run at V m/s toward the train while the train accelerates and is able to board the train some time between 6 s after the train begins to move and 10 s after the train begins to move. What is the minimum speed the commuter can have? What is the maximum speed the commuter can have?*

This is clearly a different problem although it deals with the same situational setting. The ALGEBRAIC PROPOSER allows one to reassign the values of the magnitudes of the originally entered quantities. Suppose, therefore, we reassign the value of the commuter's speed which is currently 8 m/s to be V m/s. Further in order to compute the minimum speed of the commuter, we will reassign the value of the time variable to be the maximum value it can have, i.e., 10 s. The ALGEBRAIC PROPOSER then transforms the screen so that it appears thus:

	How many	What		Notes
A	V	m/s		speed of running commuter
B	30	m		initial train-commuter separation
C	1	m/s*s		acceleration of train
D	10	s		time variable
E	V*10	m	A*D	distance of commuter from start at time T
F	10	m/s	C*D	instantaneous speed of train at time T
G	1/2			constant needed to compute average speed
H	5	m/s	G*F	average speed of train during time T
I	50	m	H*D	distance train travels in time T
J	80	m	I+B	distance of train from start at time T

The inequality to be solved is transformed so that it reads

$$V*10 > 80 \text{ m}$$

This reader will note that the referent of the inequality is meters despite the fact that the referent of the variable is meters/second.

The dramatically different appearance of this screen is to be contrasted with the screen that contains the semantic graph of the problem situation. It remains *unchanged!*

The screen containing the semantic graph of the situation describes not only these two problems that we have considered here, but rather an ensemble of possible problems that can vary as to which quantities are known by value and which quantities are variable.

The fact that the semantic graph is invariant across a wide range of problems thus leads us to an important new consideration, namely the idea of 'classes of problems' What might it mean to say that two problems belong to the 'same class of problems'?

Our analysis of the two problems above would seem to provide a reasonable initial answer to this question. Problems in which the semantic graph of the referent quantities are the same might be said to belong to the same class of problems.

We might, however, consider a more general answer to the question of what it might mean for two problems to belong to the same class of problems. Suppose we were to consider two problems as belonging to the same class of problems if their semantic graphs are topologically equivalent. The definition of problem similarity given above is clearly a special case of this more general definition.

Whether we limit ourselves to the more restricted definition of problem similarity or not, the fact remains that the software provides an opportunity for users to engage a particular mathematical construct and the software environment can suggest a class of constructs to which the user's construct belongs.

How General Is the Scheme?

Are there domains, other than that of algebraic modeling that lend themselves to the use of this sort of tutor? It seems that there may well be. Let us see if we isolate the essential structural features that made possible the tutoring strategy of the ALGEBRAIC PROPOSER.

- The elements that are subject to analysis within the domain are well defined. They can each be described in a variety of representations, but particularly symbolically and verbally.

- There is a relatively small and well defined set of operations that may be carried out on these elements.

- The transformed elements that result from the operations carried out on elements are similar enough in kind to the original elements that they can be described with the same set of representations.

What other domains might be candidates for which this sort of tutor might be devised? In particular, might school geometry be a candidate for the building of such a tutoring environment?

In the case of geometry, the primitive elements are geometric objects such as points, lines, polygons, circles, etc. Alternatively, we might take the primitive objects to be points alone. This actually conforms rather closely to the underlying data structure in the Geometric Supposer (Schwartz & Yerushalmy, 1985) and in the Geometric superSupposer (Schwartz & Yerushalmy, 1992).

As for operations, if we limit ourselves to Euclidean geometry then there are a limited number of operations that can be carried out on these objects. Guided by the primitive operations of the Geometric Supposer we consider these to include the drawing of line segments, parallels and perpendiculars, angle bisectors and circular arcs. The objects resulting from these operations can be described in the same representations as the primitive objects.

Thus it would seem that the necessary conditions cited above for the building of this sort of Socratic tutor obtain. Necessity, however, is not sufficiency. Let use consider a geometry problem with an eye toward seeing whether a semantic graph might be generated and then used by a tutor.

Given AB, CD and EF are three non-parallel line segments in the plane, draw all the circles that are tangent to the lines AB, CD and EF.

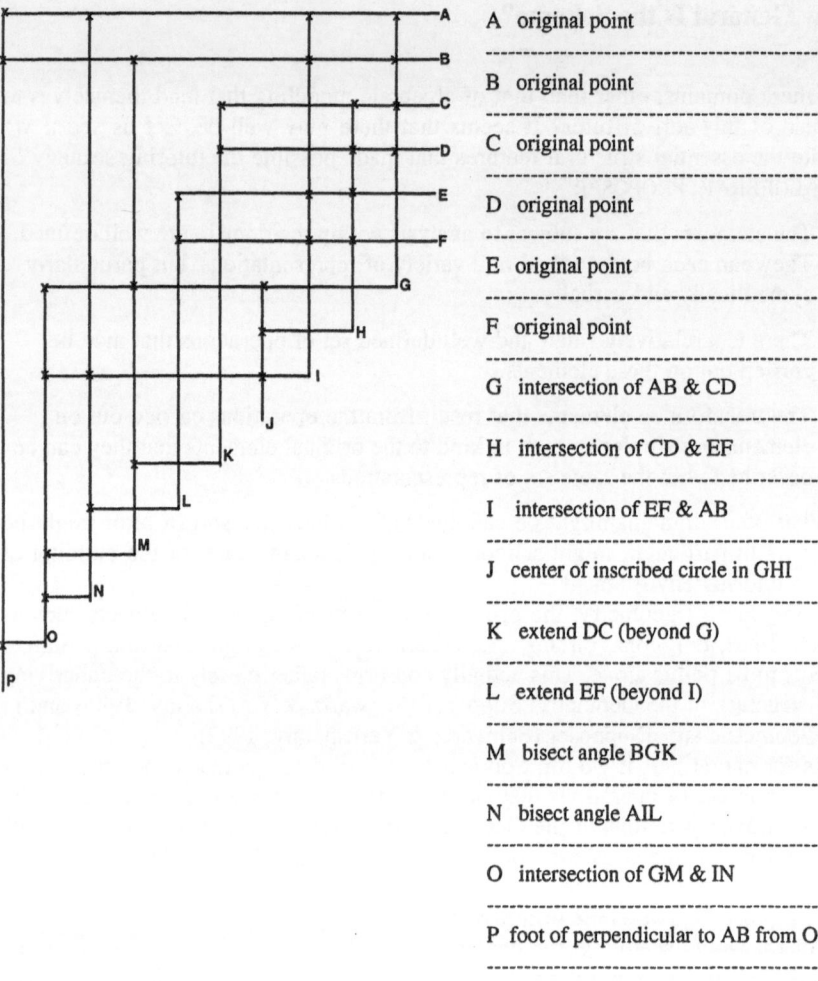

A original point

B original point

C original point

D original point

E original point

F original point

G intersection of AB & CD

H intersection of CD & EF

I   intersection of EF & AB

J center of inscribed circle in GHI

K extend DC (beyond G)

L extend EF (beyond I)

M bisect angle BGK

N bisect angle AIL

O intersection of GM & IN

P foot of perpendicular to AB from O

There are several things to be noted about this semantic graph. First of all, it might be argued that points are too primitive a structure and that the appropriate primitive element for this representation is something like a line segment e.g., AB or CD. Were it the case that all subsequent actions involved A and B symmetrically and simultaneously, this claim would deserve further consideration. Given that that is not the case, we retreat to the structure in the Geometric Supposer which is one built on points in the plane.

It can be seen from the semantic graph that the procedure to generate J and the procedure to generate P commute with one another, i.e. J could be found before P. Note also that the procedure to generate J uses the original pairs of points (A, B; C, D; E, F) symmetrically to generate the triplet of points G, H and I, which in turn are used to generate J. In contrast, the procedure to generate P uses two of the original pairs (C, D; E, F) symmetrically. This implies that it can be repeated two more times, using cyclic permutations of the three original pairs to generate yet another two circles.

We make the obvious note in passing that if the original points were paired differently, i.e., different line segments were constructed using the same points, the same set of procedures would have led to construction of a different set of tangent circles.

Does all this mean that the semantic graph representation is useful in geometry? In order to answer this question, we have to ask, useful to whom? It seems that the graph might be too complex visually to be of direct use to students. Might it be of use in a software environment designed to guide a student in a Socratic fashion through a domain, without inferring intention? We believe that there is sufficient promise in the approach to warrant further investigation.

Some Thoughts in Conclusion

Finally, one should ask a fundamental question, i.e., is the system described a tutor at all? After all, it is possible to argue that all it is doing is scaffolding the students interaction with the problem and providing a set of intellectual 'training wheels' for the problem solver. Perhaps so. But there are those that believe that students best served by tutors whose actions cause the students to find their own way through the problems they face.

I do not expect those doing research on intelligent tutoring systems to find the arguments put forward here so compelling that they will drop what they are doing to pursue the advantages and disadvantages of the perspective I have presented. Research on more traditional ICAI and computer based tutors will and should continue. Are there, however, any implications of the perspective presented in this paper for that work? I believe there are two.

From the point of view of the student user, present day ICAI systems, for the most part, are anonymous. In contrast, written curricular materials are signed by their authors. This practice permits the student, in principle, to understand that the materials are authored by a human being with a certain set of understandings and perspectives in the domain. In a more perfect educational world, we would encourage students to think for themselves with enough passion to take issue with authors with whom they disagreed. Of course we would train them to assemble powerful arguments to support their points of view, but those arguments must be addressed to someone. Anonymous materials, including anonymous ICAI systems, work at cross purposes to this end.

Related to, but logically distinct from, the problem of anonymity, is the lack of facility in most extant ICAI systems for the student to query the system about the reasons that underlie the system's decisions and reactions. For example, if on the basis of a series of interactions between the system and the student, the system recommends a particular problem for the student to try, the student ought to be able to ask 'why this problem?' In response, the system ought to expose its model of that student and the reasoning that led it to decide that the problem it presented to the student is indeed appropriate.

It seems to me that signed systems that expose their internal models to users on request is the intellectually respectable way to proceed. Moreover, given the difficulty of building semantic models of domains that incorporate nuance and subtlety, and given the fragility and oversimplification of almost all our models of learners, I think that researchers in this field are morally obligated to do so, so as to not put students and teachers at risk.

References

Schwartz, J.L. (1987). The Algebraic Proposer: A microcomputer modeling environment for analysis and problem-solving. Hanover, NH: True BASIC.

Schwartz, J.L. & Yerushalmy, M. (1983–87). The Geometric Supposer. Pleasantville, NY: Sunburst Communications.

Schwartz, J.L. & Yerushalmy, M. (1992). The Geometric superSupposer: A laboratory for conjecture, exploration and invention. Pleasantville, NY: Sunburst Communications.

Sleeman, D. and Brown, J.S. (eds.) (1982). Intelligent Tutoring Systems. Academic Press, London.

White, B.Y. and Frederikson, J.R. (1986). Intelligent Tutoring Systems based upon Qualitative Model Evolutions, Proceedings of AAAI-86: The National Conference on Artificial Intelligence, Philadelphia, PA. August 1986.

Students' Constructions and Proofs in a Computer Environment – Problems and Potentials of a Modelling Experience

Rudolf Strässer

IDM Bielefeld, Universität Bielefeld, Postfach 10031, D-33501 Bielefeld, Germany
E-mail: rstraess@hrz.uni-bielefeld.de

"I would rather ask what geometry is on the lowest, the bottom level? ... Geometry is ... grasping the space in which the child lives, breathes and moves. The space that the child must learn to know, explore, conquer in order to live, breathe and move better in it."

[Geometry auf der niedrigsten, der nullten Stufe ist ... die Erfassung des Raumes, in dem das Kind lebt, sich bewegt, den es kennen lernen muß, den es erforschen und erobern muß, um besser in ihm leben, atmen und sich bewegen zu können.]

(Freudenthal 1973, quoted from the English edition, p. 403)

"Instead of 'points', 'straight lines' and 'planes', one must always be able to speak about 'tables', 'chairs' and 'beer-mugs'."

[Man muß jederzeit an Stelle von "Punkten, Geraden, Ebenen", "Tische, Stühle, Bierseidel" sagen können.]

(Hilbert in a Berlin pub, reported in: Meschkowski 1964, p. 122; translation to English: R.S.)

Abstract. Starting from the complementarity of empirical concepts (related to the physical, material world) and theoretical (sometimes logical) concepts pertinent to geometry, the paper reports on an experience with the software tool CABRI-géomètre. The software-representation of the use of empirical and/or theoretical concepts will be described, analysed and confronted to the students' handling of the tool in a construction task. Consequences for the modelling of constructions and proofs as part of the student knowledge in geometry focus on the computer-evaluation of empirical approaches and the complementarity of the visual representation on the screen (biased to empirical concepts) and the internal representation of a solution (biased to theoretical concepts).

1. The Case of Geometry: Empirical Versus Theoretical Concepts

The quotations above show two poles of a complementary which in some sense is at the heart of geometry: For thousands of years, geometers struggled for insight into the (in)dependence of the fifth postulate of Euclid from the other postulates – with no solution by empirical findings (e.g., measurement) and no deduction from the theoretical concepts formulated by Euclid and his successors. One had to wait till Gauss, Bolyai and Lobatschewski and their non-Euclidean models to know about the logical independence of the axiom on parallel lines from the other axioms, thus raising the empirical question whether the 'real world' has a Euclidean or other structure or both. The discoveries (at the turn of the century brought into an axiomatic structure by Hilbert) turned parts of geometry into an analysis of space and plane (as is described in the Freudenthal quotation) and made geometry an empirical science – with concepts defined by measurement for quantitative concepts and identification and prototypic examples for qualitative ones. Most of the lower secondary geometry teaching still follows this line (cf. Struve 1987), taking geometry as a part of (theoretical) physics, coping with empirical constraints. Within an approach like this, application of geometrical concepts, procedures and statements can be explained rather easily: We start from empirical data and want to cope with empirical situations – no change of epistemological compartment is necessary. This type of geometry has relevant applications in planning and controlling the (artisanal and industrial) production of goods. Nowadays, the use of geometry is developing under the influence of computer related technologies like CAD, CIM and the integration of such techniques into overall computer based production management systems.

Nevertheless, the empirical approach has some limitations and presents too restricted a picture of geometry in the 20th century: it does not show the important role of geometry in the development of axiomatised mathematics, hides the "ontologically neutral" position of formal mathematics (cf. Curry 1951) and cannot explain the affinity of geometry to formal reasoning. As a theoretical science, geometrical concepts are implicitly defined by relations between primitive objects, geometry studies these relations and - as for centuries, geometry is linked to deduction and formal reasoning. Within mathematics as a discipline, this link was additionally supported by the axiomatic approach developed in the 20th century.

The axiomatic reconstruction of mathematics offered a way to cope with the epistemological problems in set theory and non-Euclidean geometry and handles the problems of communicating and teaching mathematics in a new way by presenting mathematics in a deductive, hierarchical, Bourbakist manner[1].

---

[1]Pedagogical implications can also be found in Freudenthal with arguments supporting 'local ordering', the process of axiomatisation and the disruption of the ontological binding (Freudenthal 1973, pp. 450ff). This discussion clearly shows that identifying Freudenthal's position on teaching geometry with his 'bottom level' would be incorrect. For the development of geometry as a scientific sub-discipline cf. (Klingenberg 1971).

Starting from this complementarist perspective, a valid description of major features of geometry and its teaching can take geometry as the effort to model the plane and/or spatial configurations by means of theoretical (especially: geometrical) concepts. The duality of the physical, spatial world and theoretical knowledge is mediated by means of a variety of graphical representations (e.g., figures in plane geometry, orthogonal views and perspective drawings in technical drawing, using descriptive geometry, cf. Laborde 1989, pp. 3–6). So a 'drawing' on paper (*dessin* in the sense of Parzysz) may represent the theoretical object, the 'figure' (cf. Parzysz 1988, p. 80; also: Arsac 1989). The specific way to mediate both aspects is culturally normed - with the graphical representations offering the possibility of conveying, planning and handling plane and spatial configurations as part of professional, social and cultural practices (the material, empirical aspect of geometry), while also making possible the study of logical dependencies, of deductions and the overall logical, axiomatic structure (the theoretical aspect of geometry). Apart from the graphical representations mentioned, geometry and its teaching can rely on a second system of signifiers: the natural and/or mathematical language. These two systems of signifiers can be brought together by the learner/user of geometry[2].

Modelling student knowledge in this field in some sense has to account for both aspects of geometry: the modelling of reality by means of geometry as well as the theoretical aspect. The above statements only illustrate the constant fight between "noospherians" (cf. Chevallard 1985), between educationalists, favouring the use and application of geometry in reality (in Germany, e.g., Timerding 1912) as opposed to those favouring the theoretical value of geometry.

2. Learning About Student Knowledge in a Computer Environment: A Modelling Experience with CABRI-géomètre

2.1 The Task – The Means of Solution

In February 1989, 12- to 14-year old students of a French 'collège' near Grenoble were given the task to construct a square using the software tool Cabri-gèomètre. Cabri-géomètre is a menu-driven software tool especially created for plane geometrical (Euclidean) constructions. In the experiment, it was run on a Macintosh machine, i.e., pull-down-menus and a mouse was available. The mouse in CABRI serves as selector from the menu-bar and also moves the cursor in order to locate the geometric elements (points, straight lines, circles) on the

---

[2]With such a perspective on geometry and its teaching, the distinction of empirical from theoretical concepts in the strict, epistemological sense (cf. Jahnke 1978, p. 70 ff, using concepts of Sneed 1971; also: Struve 1987, pp. 249ff) may be important for an epistemology of geometry, but – for the purpose of modelling students' knowledge in geometry – can be left aside. For modelling student knowledge, there is no need to exactly define a theory of geometry underlying the modelling of student knowledge.

screen which is taken to simulate a piece of paper (for a construction process see Sect. 2.2). CABRI offers the possibility of selecting the tools available in a specific construction by defining or changing the menu-bar before the application is run. Besides the 'usual' Macintosh items (like 'open', 'close', ...), the students were offered the geometrical tools to create geometrical objects (points, line segments, straight lines and circles) and activate constructions (a point on a predefined object, the intersection of two objects, the midpoint of a line-segment, the mid-perpendicular of a line-segment, a line parallel to a given line or segment and a line orthogonal to a given line or segment; cf. the menu-bar documented below). For the user's convenience, it was also possible to delete an object or a relation, to fix a point to an object, and to measure the length of a line segment (up to the first decimal).

drawing	construction
point	a point on an object
segment defined by 2 points	intersection of 2 objects
line defined by 2 points	midpoint of a segment
circle defined by 2 points	mid-perpendicular of a segment
	parallel line
	orthogonal line
	symmetric point

Figure 1. Menu offered for the construction task

The students started from a given line-segment AB with the square to be constructed in a way that its sides are of the length of AB. When they said to have finished the construction, they evaluated the construction by 'dragging' around one top of the quadrilateral they had constructed, thus evaluating if the figure constructed was a square 'in the sense of Cabri'. If the quadrilateral turned out to be no square, they started a new construction on a 'new' page with the line-segment AB. If the had offered a correct solution, they were given a blank 'sheet' (screen) and were asked to think of and present a different solution where the sides of the square could have arbitrary length (this was asked to get a greater variety of solutions). The constructions of the students were protocolled by means of a special protocol function built into CABRI, which in a series of files reports every operation that alters the geometric configuration on the screen (Sect. 2.2 gives examples of complete constructions reported with this function).

The task was given to four pairs of students and two students using the computer alone. (In order to shorten the description, the paper continues to describe student-solutions even when analysing the solution of a 'binom', a pair of students.) About half of the students were familiar with the Macintosh-environment in general, none of them had special experience with CABRI. The students were introduced to CABRI and the menu-items offered and an example of 'dragging' was shown. Before they started the construction, the interviewer/ supervisor (which was the author) asked them to recall the defining properties of a square.

All of them first stated the 'equal length' property (equilateracy), while the orthogonality of the sides sometimes was only mentioned after being confronted with the example of a rhombus. Minor technical difficulties with the handling of the computer and software could be solved on the spot during the construction process with the help of the interviewer.

2.2 Solutions

Three of the students gave solutions of the construction problem using perpendiculars and parallels to secure the right angles and a circle/circles to secure the equal length property (students # 2,3,6). To give an idea of this type of solution, the next page shows a series of constructions of student # 6, reported with the help of the CABRI protocol-function (for the completed construction see Fig. 2).

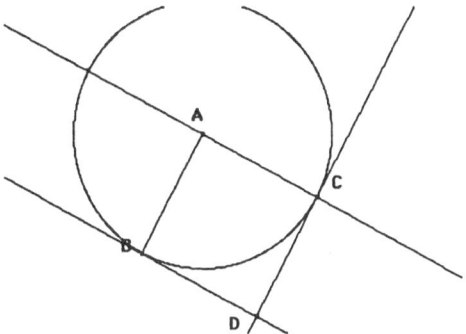

Figure 2. Construction of Student # 6

There were also more complicated correct solutions using special features of CABRI (like: point symmetry, cf. the construction of student # 3, documented in the screen dump of the end of the construction in Fig. 3) or specific properties of the square (the center of the inscribed circle is the intersection of the mid-perpendiculars of the sides of a square, cf. construction of student # 2 in Fig. 4).

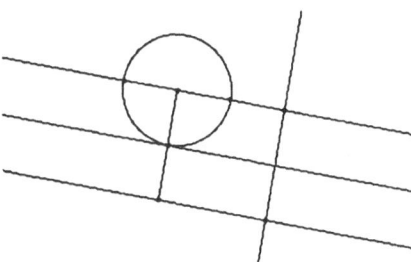

Figure 3. Construction of Student # 3

208   R. Strässer

Documentation of a construction of Student # 6

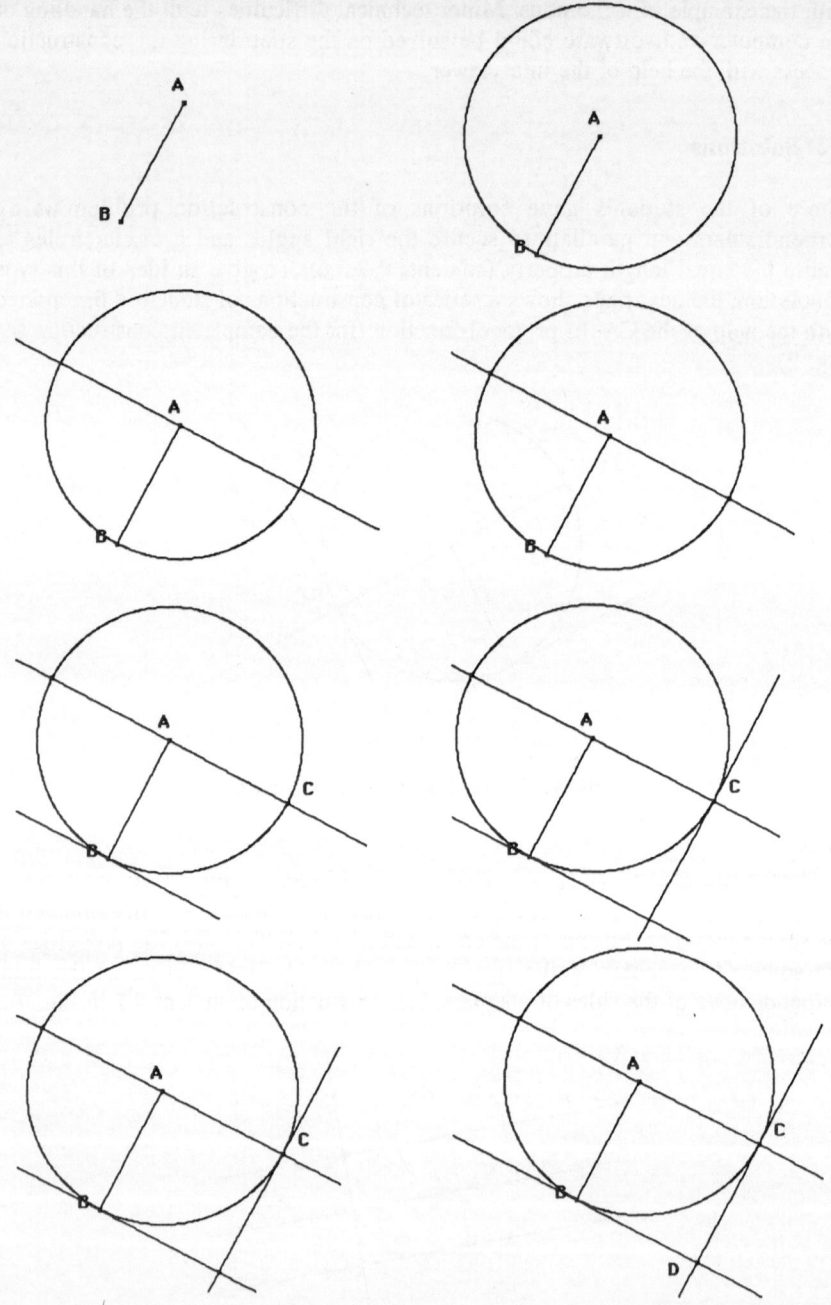

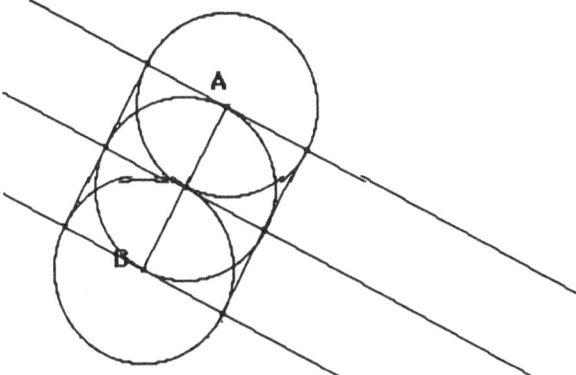

Figure 4. Construction of Student # 2

When asked for a different solution with arbritary length of the side of the square, some of the students came up with a construction using a (part of the) given side as diagonal of the square and a (mid-)perpendicular of the starting segment as a second diagonal (e.g., student # 2 with the screen dump in Fig. 5). These constructions made (sometimes sophisticated) use of geometric properties of the square and related CABRI menu-items to model these properties. Basically again, the students modelled orthogonality with the 'perpendicular' menu and transferred the length of a given segment to another by means of a circle.

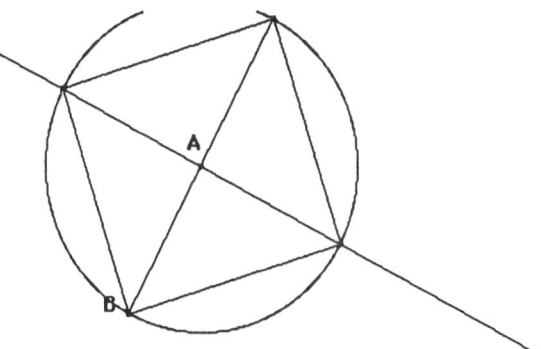

Figure 5. Construction of Student # 2

There is a second *solution* to be presented and discussed here: In their first construction, three of the students (# 3,4,5) secured the orthogonality by means of perpendiculars (and parallels) while securing the equilateracy estimating by the eye and/or numerically measuring the length of the sides and adjusting them

Documentation of a Construction of Student # 5

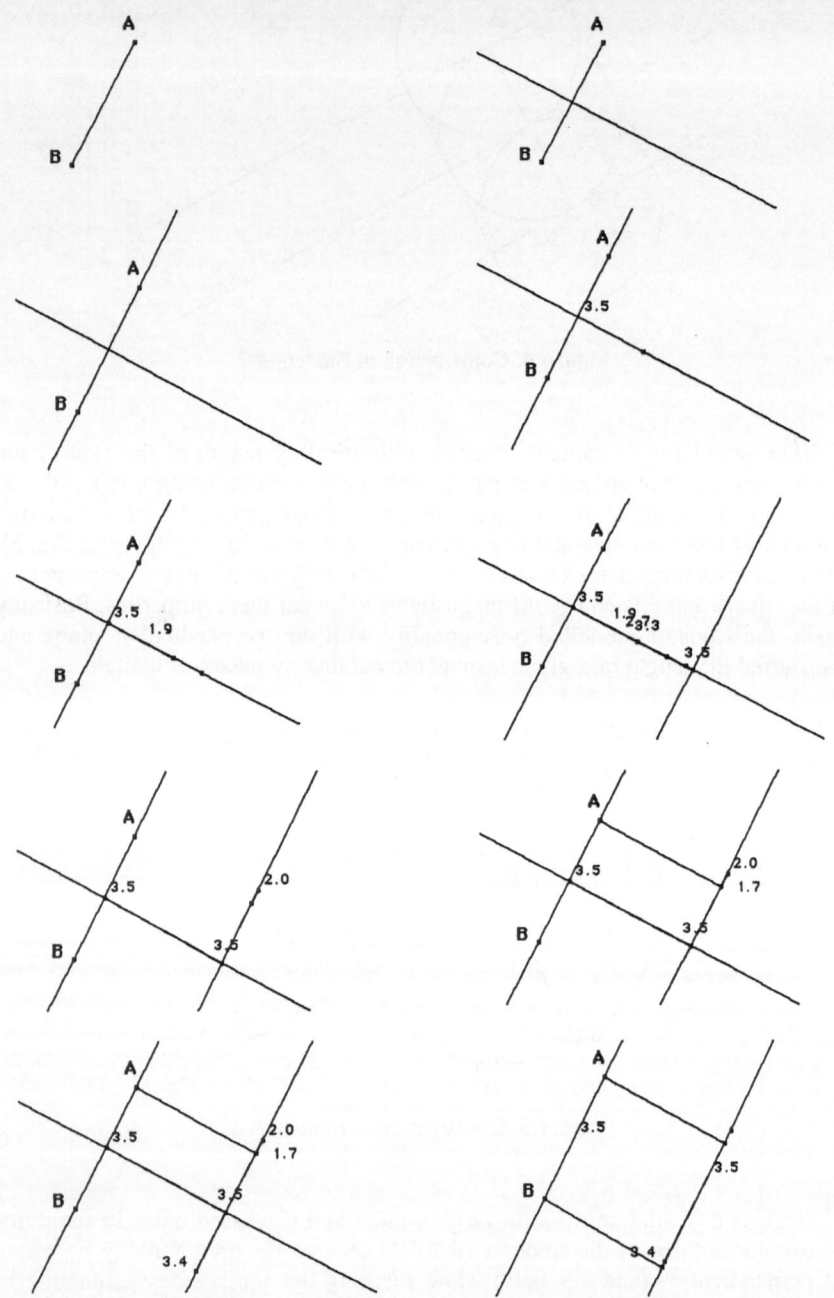

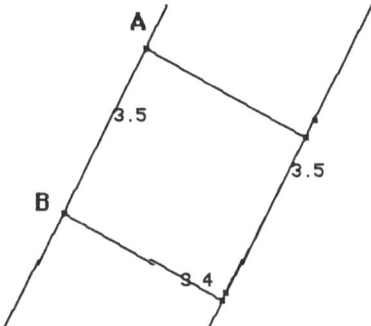

Figure 6. Construction of Student # 5

to equal value. For a purpose like this, the version of CABRI used in the experiment offers the possibility of numerically measuring the lengths of sides (rounded to the nearest millimetre). A complete construction process using this type of procedure is shown on the preceding page (for the completed figure see Fig. 6).

The student was not too happy with this solution – not least because obviously, from the measurement, one can see that one side of the quadrilateral has the wrong length.

In order to give an overall view, Fig. 7 gives a complete picture of the students' constructions including the order of the constructions – showing that during the 45-minute period of the experience, the students undertook at most four completed constructions (namely students # 2 and 3). Figure 7 deliberately omits the attempts at constructions which the students gave up during the session. (Only students # 3 and 4 gave up one construction they had already begun and asked for a new 'blank sheet of paper'. Four of them (# 1,4,5,6) were in the middle of a construction when the time allotted to the interview ran out. These constructions, which due to time constraints were not completed, are not reported in Fig. 7.)

Figure 7 shows the order of constructions offered by the students. For example, student # 2 first offered a construction which made the given line-segment AB the diagonal of a square – an incorrect construction if presented at first because of an incorrect length of the sides of the square. Then he/she constructed a square with the starting segment as one side, using a circle to secure equilateracy and perpendicular to secure orthogonality – a construction evaluated as correct. He/she then again constructed a square with the starting segment as part of the diagonal – now evaluated as correct because the square now could have sides of arbitrary length. Finally, she/he constructed a square using the inner circle of a square – see Fig. 4 for the completed construction.

Correct			Incorrect		
using inscribed circle	starting from AB as diagonal	using circle/ perpendicular	construct. rectangle	using AB as diagonal	estimation of angle by eye

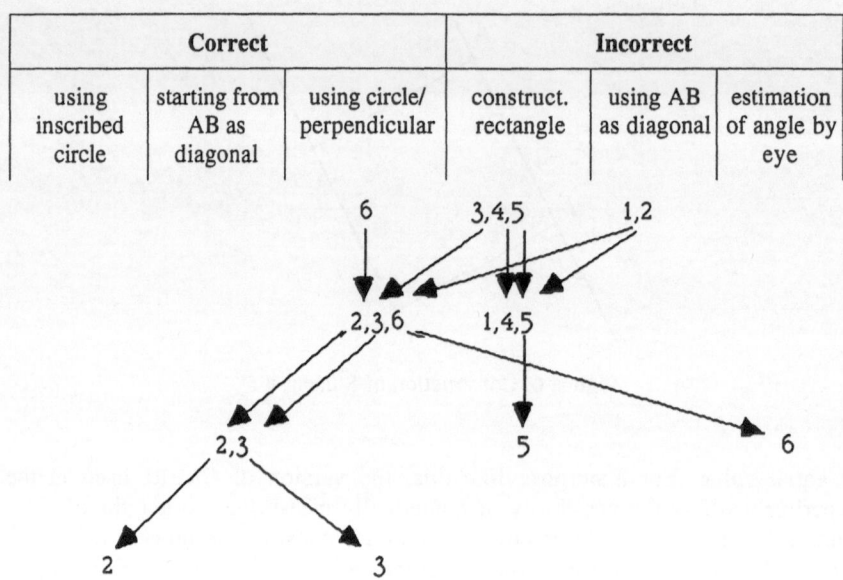

Figure 7. Solution paths

2.3 Case Studies in Empirical Versus Theoretical Constructions

If we come back to the 'equal length problem', basically three ways to secure the equal length of segments were presented: there was estimation by the eye, numerical measuring with the help of the length measurement offered by CABRI and there were constructions using a circle to secure equilateracy. During the whole interview, only student # 1 kept securing the equilateracy by estimating lengths by eye. Student # 3 started with estimation by the eye, but then turned to securing equilateracy by using circles in his follow-up constructions. Students # 4 and 5 used the length-measurement of CABRI-géomètre to have sides of equal length. In order to illustrate this method in greater detail, Figure 8 shows the end of a 'measurement-construction' of a definite length (for a whole measurement-process see part of the construction of student # 5 above).

For lengths, students # 2 and 6 during the whole interview consistently used circles to secure equilateracy – as shown for example in the construction documented in Sect. 2.2.

In the light of the distinction outlined in Sect. 1, estimation by eye as well as measuring can be classified as an empirical approach – obviously accessible to student users and of great importance when using geometry in real-life situations. The students measuring lengths in the experiment did not use the alternative construction by circles. At first, they were happy with a figure looking like a square or showing numerical values as to be satisfied by such a quadrilateral. They were happy with a visual imagery on the screen if it looked to have the desired properties of a square. The signs on the screen were taken as an object having the properties they wanted to have this object. Their happiness only was destroyed by 'dragging points' – showing that they had constructed an object which offered the picture of a square, but – 'in the sense of CABRI-géomètre' – was not a square. They then reembarked for a new construction - sometimes again using measurement to secure equilateracy. This phenomenon will be commented further down.

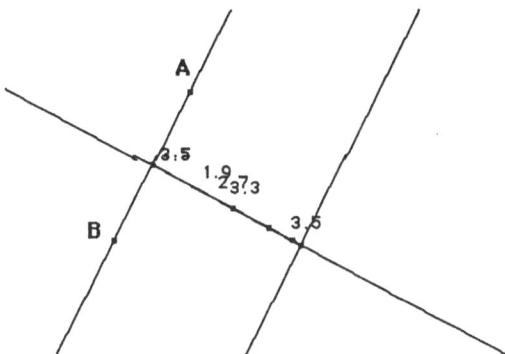

Figure 8. Measurement-construction of Student # 5

Contradictory to the feelings of the user(s), an automatised evaluation of the constructions would classify constructions by eye or by measurement as wrong, incorrect solutions of the task. (The problem of accuracy of numerical values is deliberately left out because it does not alter the underlying problem.) The internal, computer-representation of the figure will not be spoiled by the visual, empirical appearance of the object. It will be classified according to theoretical concepts used in the construction (namely equilateracy and orthogonality).

A computer model of the construction – and a model of the geometrical knowledge used or to be used – is more easily built on the theoretical characteristics of the figure - thus relying on theoretical properties of the figure.

For the modelling problem, the situation is even more complicated: CABRI in the version used in the experiment did not offer the possibility to measure angles. So all the students immediately used the menu item 'perpendicular' to secure the orthogonality – which could be surprising because most of them did not mention this property in the beginning of the experiment. The given initial segment was deliberately drawn in an oblique position to ensure that visual control of orthogonality was difficult. With only a theoretical tool offered and with no simple visual criteria of orthogonality at hand, all the students proceeded theoretically. Only student # 6, who was forced to reflect on additional and different constructions than the correct one already showed, tried to install orthogonality by estimation – with nearly horizontal and vertical lines to start from, hence a visual help. Hence, using theoretical tools does not imply the cognitive formation of theoretical concepts by the user? How to cope with such a situation in modelling student knowledge?

2.4 The Dragging Issue

To judge on and evaluate a construction, CABRI offers the possibility of 'dragging' points with the mouse, presenting continuous visual feed-back to the user/student while leaving unchanged relations enforced by the use of menu items (e.g., equal lengths when constructed via a circle). Using this feature of CABRI, the interviewer made the students drag the corners of their quadrilateral when they said the construction was finished. If the figure they had constructed did not remain a square when dragging a point, this was commented that they had not constructed a square 'in the sense of CABRI-géomètre'. All the students accepted this type of objection and set out for a new construction if time was left. Nevertheless, a certain ambiguity persisted for the students. How to interpret this type of visual information and what are the consequences of this procedure?

To use the distinction of 'drawing' and 'figure' mentioned in Sect. 1, dragging points shows a series of drawings (on the screen, not on paper) which brings to light important, theoretical aspects of the 'figure'.

To rephrase it in a more general terminology (cf. Steinbring 1989), dragging offers the possibility to visualize the difference between the 'object' constructed and the 'symbols, signs' on the screen. In contrast to the situation in traditional paper-and-pencil geometry, CABRI offers a means to distinguish the object constructed by the student, the 'figure', from the symbols on the screen, the 'drawing', by giving access to the variety of potential objects inherent in the construction of the student. Showing this variety of possible objects inherent in a construction gives access to the concepts used in the construction. The possibility of visually checking the construction by dragging helps to bridge the dichotomy of empirical and theoretical concepts, it mediates the duality of empirical and theoretical aspects of the configuration by bringing to light visually implicit properties or showing the absence of desired properties.

From discussions with the Grenoble colleagues (especially B. Capponi, who made possible the experiments in the 'collège' he is teaching in), it is obvious that 'dragging points' only offers this potential when often used, explicitly analysed and institutionalised as a means to check a construction. An explicit analysis of the dragging procedure implies the differentiation of visual representation of a configuration, the signs on the screen, the 'drawing', from the underlying relations between the elements of the construction, the 'figure'. Dragging in some sense is a way to explicitly mediate the empirical and the theoretical relations inherent in the construction. The mediation can only be used at the cost of an explicit introduction and analysis, a continuous use of this feature and an – at least implicit – distinction between empirical and theoretical concepts in a construction task.

3 Conclusion on Modelling Empirical and Theoretical Concepts in a Computer Environment

A look back on the experiment described above reveals some important features of modelling students' knowledge in geometry. They are offered in the form of statements (and some question marks in the statements).

(1) Analysing the *constructions* of plane or spatial configurations offers a viable and valuable approach to modelling students' knowledge in geometry.

(2) Modelling experiments have to consider that *modelling is a two-way activity:* The tools offered in a computer environment model the student knowledge by their mere existence - thus also enforcing the use of 'theoretical' concepts if offered exclusively. The modelling experience itself influences, if not models the knowledge a user, a student has (e.g. in geometry). This must not be taken as a disadvantage - last not least because - in other circumstances - the 'same' activity is called teaching and/or learning.

(3) The visual display of computers together with the ease of trial, error and correction in computer environments *favour empirical concepts*. The same support of empirical concepts can be found in every-day practice and teaching approaches in geometry (at least: in the lower secondary level). The problem of how to empirically evaluate a geometric construction (known to every teacher of geometry only using traditional means as paper, pencil and compass/ruler) is difficult to handle also in computer-based modelling and evaluation of students' constructions. From the product, it may be difficult to judge on the means/ concepts used by the student. Additionally, the evaluation often has to cope with a large variety of unrelated, unclassified elements.

As a consequence, computer-based modelling and evaluation may *favour theoretical concepts*. Theory-based constructions and the use of theoretical concepts are more easily evaluated automatically than empirical ones, which are difficult to catalogue in advance and evaluate in order to learn about student knowledge. As a consequence, modelling student knowledge favours and has a bias to theoretical concepts – sometimes even judging on the mere use of theoretical concepts without control on the cognitive formation of the concepts.

A comparison with computer evaluations in a LOGO environment (nearer to procedural knowledge) and/or the 'geometric suppress (closer to empirical constructions) could be helpful to clarify this point. Nevertheless, this paper does *not treat* the question of *procedural versus declarative* knowledge.

(4) As a software-tool, CABRI seems to be biased to theoretical concepts. Most of the menu-items of Cabri obviously favour theoretical concepts - and this decision was even clearer in earlier versions of CABRI. The bias is strongly supported by arguments stemming from scholarly knowledge (namely geometry as a science). A decision to narrow the distance between knowledge to be taught and scholarly knowledge as much as possible is an additional argument in the same direction. Apart from the 'epistemological vigilance' which is inherent in the fostering of theoretical concepts within CABRI, the ease to automatically check constructions in terms of relational, theoretical concepts (*not* empirically, visually) supports the predominance of theoretical concepts in this brand of

computer-based geometry. Reluctantly (??) and under heavy demands of school geometry, CABRI is now expanding to empirical concepts, e.g., by introducing angle and area measurement. Viewed as a program in geometry teaching with theoretical concepts as the main goal, the bias to theoretical concepts is justified. Nevertheless, this decision hinders the distribution and use in schools because the search for empirical representations and concepts, the pedagogical decision for Freudenthal's 'bottom-level', is predominant in the teaching system – at least in the lower secondary level.

(5) For geometry, the complementary of empirical and theoretical aspects is valuable in the a-priori analysis and a posteriority classification of a solution process. It is a major issue (and to date: an untreatable problem?) how to *model the solution process in general*. A record of activities (key strokes and screen dumps) offers a (sometimes too) rich source of insight. Today, only domain-specific approaches satisfy the need for classification and analysis.

As a final remark, there is a quotation from the famous Hilbert/Cohn-Vossen book *Geometry and the Imagination* commenting the duality of empirical and theoretical aspects in geometry:

> In mathematics, as in any scientific research, we find two tendencies present. On the one hand, the tendency toward *abstraction* seeks to crystallise the *logical* relations inherent in the maze of material that is being studied, and to correlate the material in a systematic and orderly manner. On the other hand, the tendency toward *intuitive understanding* fosters a more immediate grasp of the objects one studies, a live *rapport* with them, so to speak, which stresses the concrete meaning of their relations.
>
> As to geometry, in particular, the abstract tendency has here led to the magnificent systematic theories of Algebraic Geometry, of Riemannian Geometry, and of Topology; these theories make extensive use of abstract reasoning and symbolic calculation in the sense of algebra. Notwithstanding this, it is still true today as it ever was that *intuitive* understanding plays a major role in geometry. And such concrete intuition is of great value not only for the research worker, but also for anyone who wishes to study and appreciate the results of research in geometry.

(David Hilbert, from the preface of *Geometry and the Imagination*, June 1932, words emphasised by Hilbert)

References

Arsac, G. (1989) La construction du concept de figure chez des élèves de 12 ans. In: Artigue, M. et al. (eds.): Actes de la 13ème conférence PME. Paris, p. 85-92.

Chevallard , Y. (1985) La transposition didactique. Grenoble: Pensées sauvages.

Curry, H.B. (1951) Outlines of a Formalist Philosophy of Mathematics. Amsterdam: North-Holland (3rd ed. 1970).

Freudenthal, H. (1973) Mathematics as an Educational Task. Dordrecht: Reidel.

Hilbert, D., Cohn-Vossen, S. (1952) Anschauliche Geometrie. Berlin: Springer 1932. (Quotation from the English translation of Geometry and the Imagination. New York: Chelsea 1952).

Jahnke, H.N. (1978): Zum Verhältnis von Wissensentwicklung und Begründung in der Mathematik – Beweisen als didaktisches Problem. Bielefeld: IDM (Materialien und Studien vol. 10).

Klingenberg, W. (1971) Grundlagen der Geometrie. Mannheim: Bibliographisches Institut.

Laborde, C. (1989) L'enseignement de la géométrie en tant que terrain d'exploration de phénomènes didactiques. (Cours à la 5ème école d'été de didactique des mathématiques et de l'informatique, Sept. 1989). Recherches en Didactique des Mathématiques, 9.3, 337-364.

Meschkowski, H. (1964) Mathematiker-Lexikon. Mannheim: Bibliographisches Institut.

Parzysz, B. (1988) Knowing vs. Seeing. Problems of the plane representation of space geometry figures. Educ. Studies in Mathematics, 19.1, 79-92.

Sneed, J.D. (1971) The logical structure of mathematical physics. Dodrecht: Reidel.

Steinbring, H.(1989) Routine and Meaning in the Mathematics Classroom. For the Learning of Mathematics, 9,1, 24-33.

Struve, H. (1987) Eine Analyse des begrifflichen Aufbaus der Schulgeometrie als Grundlage einer Didaktik der Schulgeometrie. Köln: Habilitationsschrift.

Timerding, H.E. (1912) Die Erziehung der Anschauung. Leipzig: Teubner.

Some Hyperbolic Geometry with CABRI-Géomètre

Marie-France Thibault and Robert La Barre

Département de mathématiques et d'informatique
Université du Québec à Trois-Rivières, Trois-Rivières, Québec, Canada
E-mail : Marie-France_Thibault@uqtr.uquebec.ca
         Robert_Labarre@uqtr.uquebec.ca

Summary. We consider the teaching of Euclidean and non-Euclidean geometries to students enrolled in a university program in Mathematics Education. We point out several problems encountered by students during this teaching which provide information about student's knowledge in geometry. We then propose a partial solution to these problems by using the CABRI-Géomètre software to illustrate several properties of hyperbolic geometry. Finally, we discuss the interest of this type hyperbolic geometry software and we propose desirable features for such a software.

Introduction

The teaching of geometry at university level makes it possible to discover a number of findings on conceptions or misconceptions that students have about fundamental notions of geometry. These conceptions have been acquired at the high school level, which corresponds to grades 7-11 in the province of Québec in Canada. These conceptions might be correct in the context of Euclidean geometry; the problem comes from the fact that they tend to become definitive and absolute in the students' mind. Another problem is the fact that students do not realize the necessity to have axioms in order to prove results.

First, we depict students enrolled in our programs. Next, we present the objectives and the pedagogical approach for the university level course we teach. This makes it easier to understand the context in which we identified a number of misconceptions or difficulties students have about geometry. Finally, we discuss how it is possible to use a software like CABRI-Géomètre [1] to overcome several of these difficulties by illustrating fundamental concepts of hyperbolic geometry. Moreover, it seems interesting to consider the development of a CABRI-Géomètre–like software with the particular specifications of the hyperbolic geometry. This new software could use the Beltrami-Klein disk model and the Poincaré disk model and should not be limited to the Euclidean model as is CABRI-Géomètre in its current version.

Students' Knowledge and Skills

The teaching of geometry in the schools of the Canadian province of Québec has changed in the last twenty years. Until the 1970s, Euclidean geometry was taught. Next, for a brief period, the geometry has almost disappeared from high-school curriculum. Now, transformation geometry is being taught to students between 12 and 16 years of age. The main objective of teaching geometry is "to help students analyze geometrical situations" [2]. In fact, students get acquainted with the vocabulary of transformation geometry : transformation, rotation, reflection, dilation and so on. They have to perform simple constructions : similar triangles, similar figures, comparison of polygons, etc. However, students are not required to prove or demonstrate results and they do not learn very much about the important results of the classical geometry.

An outcome of this situation is that students at university level are familiar with the vocabulary of geometry and they can solve problems in R^2 and R^3. However, since they have learned geometrical concepts concurrently with concepts from calculus, analytical geometry and trigonometry, their geometry vocabulary does not constitute a consistent set of notions really distinct from the other parts of mathematics. Finally, students know few of the main classical results of geometry.

At university level, the geometry course is attended by students from the Bachelor degree in Mathematics and the Bachelor degree in Mathematics Education (high school). It is given at the beginning of the second year of a three-year curriculum. In the first year of the program, students develop the ability, amongst other things, to prove theorems. The geometry course aims more specifically at the development of a deep understanding of the fundamental concepts of geometry; developing skills in making constructions or manipulations is secondary. A specific learning objective is to construct an axiomatic system for the plane geometry. Given students' skills and knowledge in mathematics, we think this is a reasonable objective.

Course Organization

M. J. Greenberg's book *Euclidean and Non-Euclidean Geometries, development and history* [3] is the reference book for the geometry course.

At the beginning of the course, Hilbert's axiomatic theory of the plane is presented, this theory being a modern version of Euclid's axiomatic theory. The introduction of parallelism axioms is postponed until the results of the neutral or absolute geometry are considered. These results are relevant in both hyperbolic and Euclidean geometries.

First, the notions of point and line are presented and the incidence axioms are introduced. Several models are developed : amongst them, finite geometries and R^2. It is also interesting to introduce at this moment the Beltrami-Klein disk model and the Poincaré disk model from the hyperbolic geometry. Then, the notion of parallel lines is defined, two lines being parallel if no point lies on both of them. Geometrical models can have different properties with regard to parallel lines:

- a model has the elliptic parallel property when no parallel lines exists in the model;
- a model has the Euclidean parallel property if for every line and every point not on the line, there exists one and only one line through the point parallel to the line;
- a model has the hyperbolic parallel property if for every line and every point not on the line, there exists at least two lines through the point parallel to the line.

The axioms of betweenness, the axioms of congruence for segments and for angles, and the axioms of continuity are introduced. The main results of neutral geometry are presented.

Then, Euclid's axiom of parallelism, its main equivalent forms and its consequences are presented. Negation of Euclid's axiom leads to the hyperbolic geometry, a theory which is as consistent as the Euclidean one.

Problems in the Course

Here is a list of the main difficulties encountered by students in the course described above; they enable one to identify several of students' misconceptions about geometry.

Unquestioning adhesion to the Euclidean model

During their previous studies, students assimilated, sometimes correctly, some-times incorrectly, a number of concepts and results on geometry. One of these concepts is the Euclidean vision of the plane and of the space with respect to parallel lines. Even though they do not know how to prove or verify this vision, students are convinced that this is 'the representation' of the world. Indeed, this representation appears in almost every model they construct and is used in many arguments of their proofs. Since it is the only representation they have been exposed to, it constitutes the only model of geometry. It is very hard for them to question this conception. After a while, they agree to develop other models but they continue to use equivalent forms of the Euclidean axiom in sketches of their proofs. They do not realize that these forms are equivalent to the Euclidean axiom and questioning this axiom does not imply for them questioning its equivalent forms.

Difficulty in perceiving the relevance of axioms

The need for axioms is not obvious for students. The necessity of an axiomatic method is not assimilated. For example, the elementary continuity principle, which says that a line segment with an endpoint inside a circle and the other endpoint outside has to intersect the circle, is obvious for students as it was at

Euclid's time. For them, there is no way to do otherwise and it is not necessary to emphasize this fact. They have also been exposed implicitly to many results and they assimilated them unconsciously. For instance, a line is infinite, there is always a point between two points and so on. Students do not understand why they have to make axioms with these statements.

Difficulty with the geometry vocabulary

There is often confusion between the concept represented by a word within the boundaries of a given domain and the everyday meaning of the word. For instance, we use the French word *droite* to denote the geometrical 'line'; but the usual meaning of the word *droite* is a straight line. Then, for students, a line is always a straight line and there is no need to verify it; moreover, a line cannot be a curved line. So there is confusion between the significant and the meaning. Another example is the notion of a triangle, the triangle being formed by three noncollinear points and the three line segments generated by these three points. Students often mix up this notion of geometrical triangle with the triangle manipulated by children, that is, the one that includes the interior of the triangle. The set theory word 'contained' is confused with the usual meaning of 'is inside': the visual perception of 'is inside' leads the students to skip the verification step in which it must be shown that every point has to satisfy the property determining the belonging to the set. The wrong interpretation of this word appears with concepts like the convexity of sets.

Difficulties about figures

The use of geometrical figures generates two kinds of difficulties. First, there is the difficulty of imagining more than one figure satisfying given assumptions. Once such a figure has been drawn, the figure influences subsequent trials to find other figures sufficiently different from the first one. Students often have several figures, all of them belonging to the same class with respect to some specification. They use features of these figures to solve the problem, neglecting other possible figures satisfying the same set of assumptions. An example is the exclusive consideration of convex quadrilateral when one has to work with quadrilateral. In this case, students do not realize the necessity to check if the quadrilateral has to be convex; they use properties of convex quadrilateral even though the quadrilateral does not have to be convex to satisfy the assumptions.

Another problem arises when there really is only one class of figures satisfying a given set of assumptions. In this case, the figure's inherent properties are often perceived by students as obvious and they do not see what must be demonstrated in order to prove the properties. For instance, when students begin to work with the notion of Saccheri quadrilateral (a quadrilateral whose base angles are right angles and whose base-adjacent sides are congruent to each other), they draw automatically a convex quadrilateral and they do not think that it could be different. So, they do not understand why it is necessary to verify that a Saccheri quadrilateral is really convex.

Difficulty in distinguishing between what is known and what can be used in proofs

It might be advantageous to construct a theory when several results of this theory are already known by students. However this can become a drawback when students mix up results already proven in the theory with results which were previously known but which were not yet proven. During the course, students often emphasize the trouble they have to distinguish between these two classes of results.

Illustrating Properties of Hyperbolic Geometry with the CABRI-Géomètre Software

In order to help students to overcome the difficulties we presented above, let us see how manipulation of hyperbolic geometry models can help students to become aware of their geometrical conceptions and to make them evolve whenever appropriate.

In a first step, let us see potential manipulations. To illustrate them, we use the CABRI-Géomètre software; this software was developed as an help to teach geometry by the 'Laboratoire de Structures Discrètes et de Didactique' of the University of Grenoble in France. Amongst other things, it makes it possible to construct figures from their specifications, for instance, triangle, perpendicular bisector, etc. One can modify figures in real time by dragging one of their base elements (point, line and so on) without losing their properties. In this way, the software provides several models for the same geometrical form.

Even if it has not been developed as a tool for teaching hyperbolic geometry, CABRI-Géomètre enables one to illustrate several of hyperbolic geometry properties. Below, we use the Beltrami-Klein disk model and the Poincaré disk model.

For the Beltrami-Klein model, we fix once and for all a circle C in the Euclidean plane. The points of the hyperbolic plane are the points in the interior of the circle C and the lines of the hyperbolic plane are the open chords (line segments without their endpoints) in the interior of C.

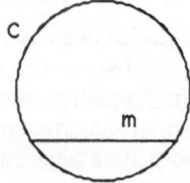

Figure 1. Line in the Beltrami-Klein model

In Figure 2, it is easy to see that this model has the hyperbolic parallel property, i.e. that for every line m and every point P not on this line, at least two distinct lines, parallel to the original line m, pass through the point P.

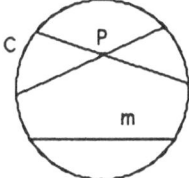

Figure 2. The hyperbolic parallel property of the Beltrami-Klein model

In the Poincaré disk model, a fixed Euclidean circle D is also chosen. The hyperbolic points are those which are in the interior of the circle. There are two kinds of lines. Open chords which pass through the center O of the circle D represent the first type of lines ; they are called diameter lines. The other lines are represented by open arcs of circles orthogonal to the circle D which are inside the circle; they are called arc lines.

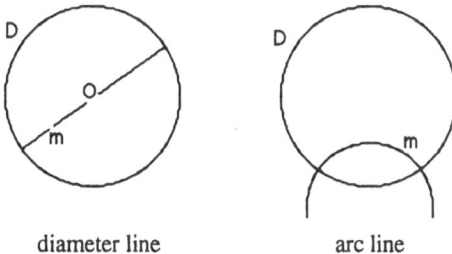

diameter line   arc line

Figure 3. Lines in the Poincaré disk model

(In its current version, CABRI-Géomètre does not allow one to erase the part of orthogonal circles to D which is on the outside of the circle D.)

In Figure 4, we can also visualize the existence of more than one parallel line to a line m passing through a point P in this second model.

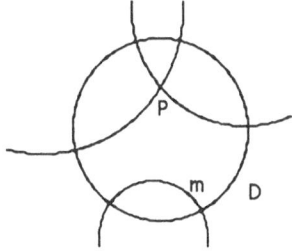

Figure 4. The hyperbolic parallel property of the Poincaré disk model

In the Poincaré disk model, congruence for angles has the usual meaning. The number of degrees in the angle generated by two directed arc lines m and n which intersect each other at a point A is the number of degrees in the angle between their tangent rays tm and tn at A. If one of the lines (or both of them) is a diameter line, we calculate the angle with this (or these) ordinary ray(s) rather than the tangent ray(s).

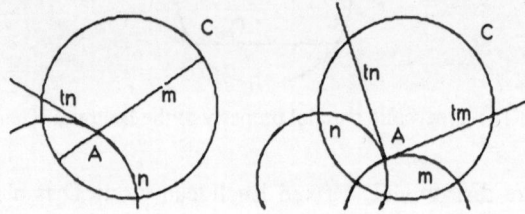

Figure 5. Degree measures of angles in the Poincaré disk model

The user can also construct triangles and study angles. The following figure shows that the sum of the degree measures of the three angles of a triangle in this model of hyperbolic geometry is less than 180°.

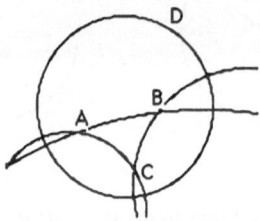

Figure 6. Triangle in the Poincaré disk model

The user can also realize that it seems to be impossible to construct a rectangle, a rectangle being defined as a quadrilateral with four right angles.

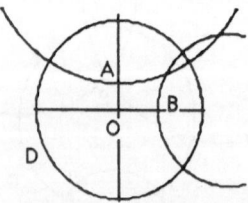

Figure 7. No rectangle exists in the Poincaré disk model

By dragging the point A and the point B in the figure above, one can verify that no rectangle with two perpendicular diameter lines exists.

Even if the Beltrami-Klein model is not conformal, congruence of angles being interpreted differently from the usual Euclidean way, we can describe the right angles or the perpendicular lines. If one of the two lines is a diameter line, the other line is perpendicular to the first line if the angle between these two lines is a Euclidean right angle.

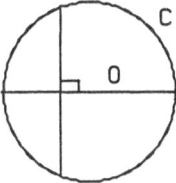

Figure 8. Perpendicular line to a diameter line in the Beltrami-Klein model

Every line m which is not a diameter line has a pole Pm, defined as the point of intersection of the two tangent lines to the circle C at the endpoints of the line, as illustrated in Figure 9.

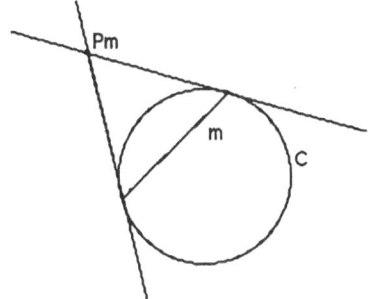

Figure 9. Pole of a line in the Beltrami-Klein model

Then, a line s is perpendicular to another line m different from a diameter line if the Euclidean extension of the line s passes through the pole Pm of the line m, as shown in Figure 10.

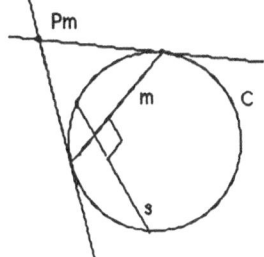

Figure 10. Perpendicular line to a non-diameter line in the Beltrami-Klein model

This representation of the perpendicularity in the Beltrami-Klein model makes it possible to discover two kinds of parallel lines in hyperbolic geometry.

In the next figure, the parallel lines m and n have a common perpendicular line s (this is always true in Euclidean geometry) only if the Euclidean line passing through the poles Pm and Pn of the lines intersects the circle C.

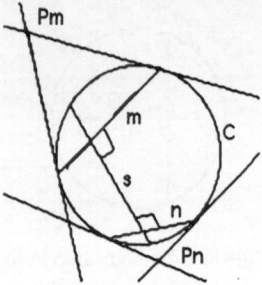

Figure 11. Common perpendicular line to two lines in the Beltrami-Klein model

If lines m and n have a common ideal point, i.e., a common point on the circle as shown below, the Euclidean line passing through the poles Pm and Pn of the lines passes through this ideal point and does not go into the circle: so, such lines cannot have a common perpendicular line.

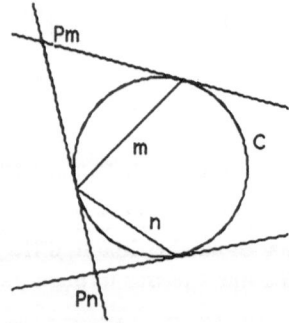

Figure 12. No common perpendicular line to two Beltrami-Klein lines
with a common ideal point

Thus, there are two kinds of parallel lines in the Beltrami-Klein model, the asymptotic parallel lines (sharing a common ideal point) and the other ones, the divergently parallel lines, also called ultraparallel or hyperparallel lines, as shown in Figure 13.

We also find these two kinds of parallel lines in the Poincaré disk model, as shown in Figure 14.

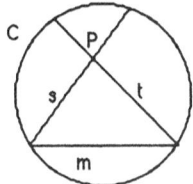
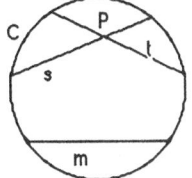

asymptotic parallel lines to m through P   divergently parallel lines to m through P

Figure 13. Parallel lines in the Beltrami-Klein model

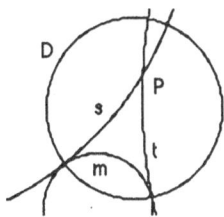
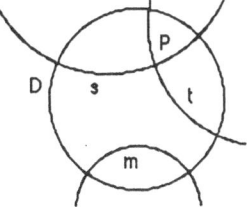

asymptotic parallel lines to m through P   divergently parallel lines to m through P

Figure 14. Parallel lines in the Poincaré disk model

In Figure 15, there is an instance of a common perpendicular line s to two divergently parallel lines m and n in the Poincaré disk model.

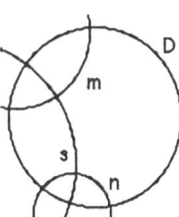

Figure 15. Common perpendicular line to two divergently parallel lines
in the Poincaré disk model

With the representation of the perpendicularly in the Beltrami-Klein model, it is easy to understand why it is impossible to construct rectangles in this model. Indeed, two asymptotic parallel lines have no common perpendicular line. Two divergently parallel lines can have only one common perpendicular line because the Euclidean extension of the perpendicular line must pass through the poles of the two lines and is completely defined by these two points. It is thus not possible for two parallel lines to have two distinct common perpendicular lines and, consequently, rectangles cannot exist in this model.

Pedagogical Aspects of Using a CABRI-Géomètre-like Software for Teaching Geometry

Even if CABRI-Géomètre is used in a context different from its original purpose, it maintains one of its important features which is to free geometrical form from its drawing constraints. In a short time, it gives many instances of a geometrical concept and one can visualize what is specific to the form and what is specific to the figure. So CABRI-Géomètre can help students with regard to the difficulties about figures as described above.

Working with the Poincaré disk model, where lines are not straight lines, helps students to question their visual perception of things. What is seen is not necessarily true or it is not the only representation of reality. Students can better discover the need to have formal proofs rather than simply to rely upon visual perception or drawing.

Working with lines which are not straight lines also helps students to grasp the difference between a geometrical concept and the literal meaning of the word used to describe it. A geometrical statement says something about geometrical concepts and not about the reality expressed by the literal meaning of the words. It is possible that the statement says something about the physical world but one must pay attention to its geometrical reality. Manipulating hyperbolic models helps solving the difficulty about vocabulary.

Moreover, the hyperbolic models give to students models different from the Euclidean one. It disturbs students' adhesion to the Euclidean plane as being the only representation of reality. Students learn to think about other models. Sometimes, the Euclidean model of geometry shows so well properties and these are so obvious to students that they do not understand the necessity to verify them and even, to question them. If they can manipulate unusual geometrical models, they will discover that several properties are not true in every model. Consequently, they will question themselves about what they know or what they think they know. They will also develop a better perception of the necessity to formalize concepts and, then, to use an axiomatic method.

Therefore, manipulating non-Euclidean geometrical models helps students to overcome several difficulties they have with geometry.

Interests for a CABRI-Géomètre-like Software Modeling Hyperbolic Geometry

It follows from this study that it would be interesting to have a software where models of hyperbolic geometry can easily be represented on a computer screen. Such a software should have the following features.

Thus far, we considered two models of hyperbolic geometry, the Beltrami-Klein disk model and the Poincaré disk model. Since several properties are easier to illustrate in one model while others are more intuitive in the other one, it would be advantageous to be able to show properties in either model, according to each hyperbolic property. There also exists a third classical model of hyperbolic geometry, the Poincaré half-plane model. The software should allow one to display these three models.

As the three above-mentioned models are isomorphic, another convenient feature would be to provide the facility to display sometimes one model, sometimes two of the models simultaneously. In the later case, creations and modifications performed on one model should also be performed immediately on the other model so that an user could see and analyze a concept or a property. In this way, an abstract notion will gain more independence with respect to its representation in a model.

Since many properties of hyperbolic geometry are also valid in Euclidean geometry, it would be interesting to have the Euclidean plane in parallel with one of the hyperbolic models. This makes it possible for students to visualize many properties of geometry which are not tied to the Euclidean model, that is for which the axiom of parallelism is irrelevant.

As CABRI-Géomètre offers the possibility to create objects and to make constructions, the software should make it possible to create 'points', 'lines', 'line segments' and 'triangles' ; the constructions 'point on object', 'intersection of two objects', 'midpoint', 'perpendicular bisector' and 'perpendicular line' should be available in every model. Moreover, it should be possible to construct an 'asymptotic parallel line' or a 'divergently parallel line' to a line (or to a line segment) through a point and the 'common perpendicular line' to two divergently parallel lines.

Conclusions

In order to favor a better learning of geometry concepts for the students, it is necessary to question and challenge their geometrical conceptions by the manipulation of different models. Paper and pencil might be helpful but are not sufficient for two reasons: it is not possible to generate a sufficient number of figures to fully illustrate the property being explored; moreover, the drawings are more tied to the mental image that students have of them rather than to the geometrical properties specifying them. Software like CABRI-Géomètre overcomes these limitations and allows practical work which supports steps of logical thinking in geometry.

In its present state, CABRI-Géomètre offers facilities to illustrate several properties of hyperbolic geometry. A further step would be to redesign CABRI-Géomètre so that it could work simultaneously with more than one model or to develop a CABRI-Géomètre-like software in which the parallelism property would not have been predetermined.

References

[1] Baulac Y., Bellemain F. et Laborde J.-M. (1988). CABRI-Géomètre, un cahier de brouillon informatique pour l'apprentissage de la géométrie, Manuel de l'utilisateur, Les Editions Nathan. (In the USA Brooks/Cole, in Germany Comet, in Italy Loescher Editor)

[2] Programme d'études, secondaire, mathématique, second cycle. Direction générale du développement pédagogique, Direction de la formation générale, Ministère de l'éducation du Québec, 1984.

[3] M. J. Greenberg, (1980). Euclidean and Non-Euclidean Geometries, development and history. W. H. Freeman, San Francisco.

Micro-Robots as a Source of Motivation for Geometry

Martial Vivet

Laboratoire informatique, Université du Maine
BP 535, F-72017 Le Mans cedex, France
E-mail: martial@lium.univ-lemans.fr

Abstract. We describe some robots based micro-worlds and we show how they can be useful to present some geometrical problems. The basic ideas about modelling, the need for formal systems can be examined here. The different types of models: 'models by physicists' based upon laws built from observation and measures or 'models by mathematicians' based upon geometrical descriptions needing the choice of representations can be presented. The roles of a model to represent a real world are underlined and this can lead to the necessity of proving properties in the geometrical descriptions of a given robot. The work done shows that we can find tools to motivate children for geometry, to exhibit the need for formal descriptions to be able to predict phenomena, events,... The need for proofs in geometry can be shown in very motivating environments based upon trucks, cranes, arms,... We know that the perception of the needed for a proof is crucial for beginners. The micro-worlds we have designed appear to us as a required pre-culture useful to reach better basic attitudes in geometry. In a second part, we describe the work we have done to design a shell to write intelligent tutoring systems (ITS).

The chosen domain concerns algebraic manipulation and no work has been done for geometry with it. About student modelling, we wonder about the possibilities to take account of the needed motivation. This can be an important component if we want to overpass the usual models (overlay, 'buggy rules',...). As a conclusion we try to discuss the possibility of writing ITS driving external micro-robots to teach motivating activities in geometry.

Introduction

The paper describes learning environments based on driving of micro-robots. The tools are designed to make younger children learn about new technologies. They can constitute 'real worlds' useful for introduction of 'powerful ideas' about modelling and geometry. They can offer to the learner an environment helping him (or her) to discover the role of models, the need for proving, the place of theorems as active tools to predict events or phenomena, limits in movments,... They can offer to beginners a motivation for activity in geometry. The place of (ITS) in geometry can be discussed here; probably we can do very few if the learner is not motivated enough. We feel that the problem of student modelling have to include these aspects. We don't have to reduce ourselves to representation of known or buggy theorems,... Basic questions can be set here as '*Why* learn geometry?', '*Which* role for geometry?', '*What* to learn in geometry?'

Micro-worlds to Give Motivation to Learn Geometry

We are designing micro-worlds in the Papert's [Papert 80] meaning to train young children (11–14 years old) with concepts in technology; The work is here more concerned with an approach of mechanics, electronics, programming. We use micro-robots as central tools to design an ordered set of physical devices in order to reach an ordered set of concepts, abilities or attitudes needed into the physical and technical world. The focus for learning can be driven by mastering of problem solving abilities or technical contents acquisition possible on tools involved. Activities in geometry can be introduced when needs for modelling appear to be effective to predict movments, verify, compute trajectories of points.

Our conclusion is that when this global effort is done we can reach very rich situations from the pedagogical point of view, more specifically, situations which can motivate for geometry can be reached.

1. Micro-Robots and Geometry

1.1 CARISTO: a first robot to leave the floor and reach the third dimension...

CARISTO is the name we gave to a first robot derived from the floor turtle by Papert. The shell of the turtle has only been changed by the coachbuilding of a fork-lift truck, the pencil by a fork. Theses changes, are shown in fig.1a and allow a first extension of the 2D-space available with the turtle: we have to manage the level of the fork.

Figure 1a. CARISTO

Projects like building stacks with boxes, carrying stacks are thus possible. It is possible to design algorithms to put stacks side by side according to the sizes of the boxes,... The learner can also reach possibilities to drive the truck with local impossibilities (bridge allowing only passages with the truck empty or the roof not high enough,...). With exactly the same approach as the work being done with the turtle we reach a more open problem space (Fig. 1b) and we have an opportunity to leave the plane geometry. Modelling deals here with modelling of the space where the truck runs, rectangles or squares as carried boxes, limits reached with a stack of boxes when the fork is up.

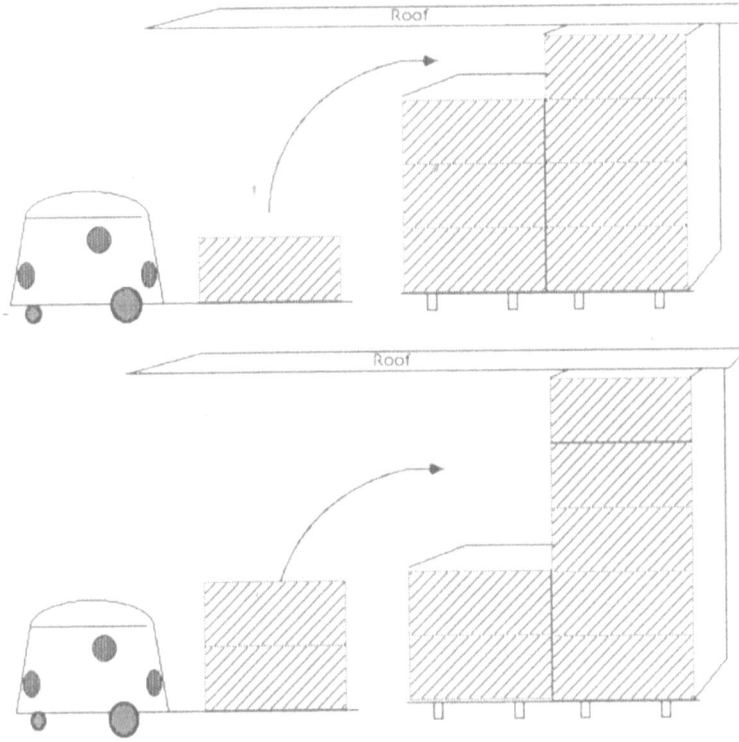

Figure 1b. Working CARISTO

2.2 CRANES: robots to solve problems in the third dimension

We have built interfaces and designed LOGO primitive commands to drive jib cranes like that shown in Fig. 2a (generally sold as toys). The language used here looks like ROTATE-LEFT, ROTATE-RIGHT, PUSH-TRUCK, PULL-TRUCK, HOOK-UP, HOOK-DOWN,... We thus reach a new space with three dimensions,

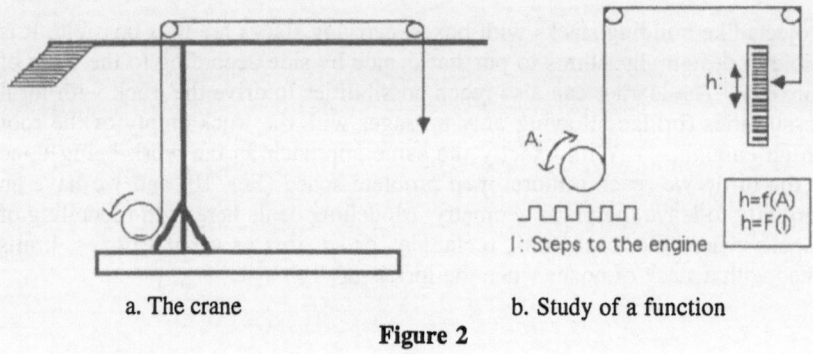

a. The crane   b. Study of a function

Figure 2

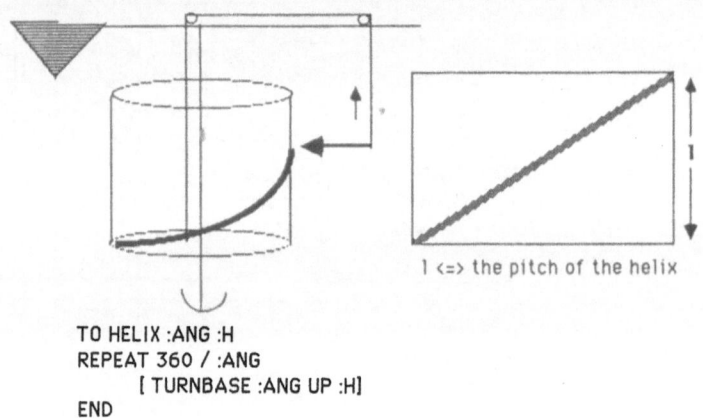

1 <=> the pitch of the helix

```
TO HELIX :ANG :H
REPEAT 360 / :ANG
        [ TURNBASE :ANG UP :H]
END
```

Figure 3. From the crane to the helix

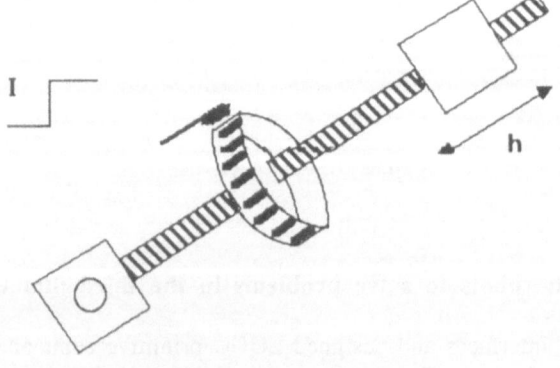

Figure 4. The screwed bar

with cylindrical and/or polar geometry aspects. The main interest here is to reach a robot with a cylindrical geometry fitting well with the geometry of the human body. So, the possibility of mental body projection available for the turtle still works and is very useful to help solving problems. This micro-world allows problems like carrying boxes, stack of boxes from a given point to a target point with possible obstacles.

A simple and realistic approach of mathematical concepts can be available with this kind of tool:

- First example: the concept of function can be studied from the study of the level reached by the hook according to the number of steps given to the engine and the radius of a winding drum. For a given radius, we can experimentally verify the linearity of the function l(s) (level reached according to the number of steps sent to the engine) as shown in Fig. 2b.

- Second example: Space modelling can be useful here: prediction in a model of the crane, limits in possible movments (external / outer cylinder). Richer problems can be found by making two or more cranes cooperate (inter-section of cylinders, circles...).

With the crane, we can also experiment and study some aspects of the geometry of the helix: the concept of 'pitch of a helix' being seen as the level reached by the hook when the arrow of the crane rotates exactly 360°, the hook moving up with a constant speed (Fig. 3). This can be a good approach for the study of screwed rods, useful for later used robots. Such rods are available to transform the rotation of the rod into the translation of a screw and can be studied as a primitive mechanical function. We have described such a work in more detail in [VIVET 86], which includes some interesting illustrations.

1.3 ARMS: robots to be built by the learners

We use Fischer-Technik toys. The main interest is in the possibility to make the pupil build, with his (her) own hands, the needed mechanics. It is a way to approach primitive mechanical functions offered by tools like gear wheels, screwed rods, crank arms,... Making an engine turn can also be studied. We have designed now LOGO primitive commands for the toys available and this appears as very interesting. Children are very happy when driving something they have built by themselves. Among other possibilities, we can give a model and a plan of a toy to be built. The children must read carefully a plan, follow very precisely a check list to reach success. This can help him (or her) in perception of spatial pictures.

With the robots sold by Fischer-Technik, an approach of geometry ('working geometry') is possible. It is possible to model precisely as shown in Fig. 5 the device and show in a fine way the role of modelling devoted to geometry, the differences between a real world and a model,... The geometry of triangles, manipulation of proportionality, properties of the parallelogramm can be studied while observing the model and the change of its shape when moving an arm (Fig. 6a,b). The study and a real use of Thales' theorem is also possible. In fact, we have observed how very important it is to have on the same desk, the model and the real world when approaching modelling activities. The figures show the relevant geometrical aspects.

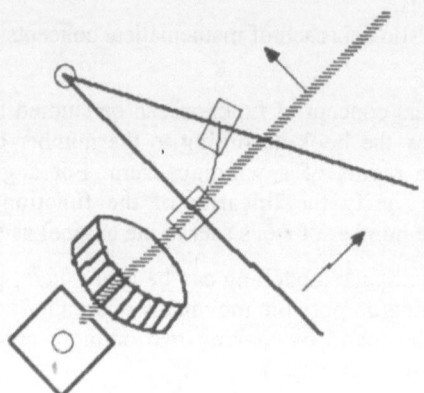

Figure 5. Moving a bar

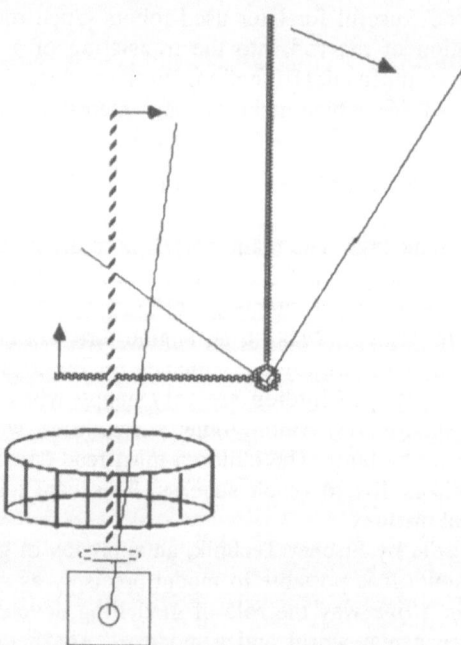

Figure 6a. Moving the arm

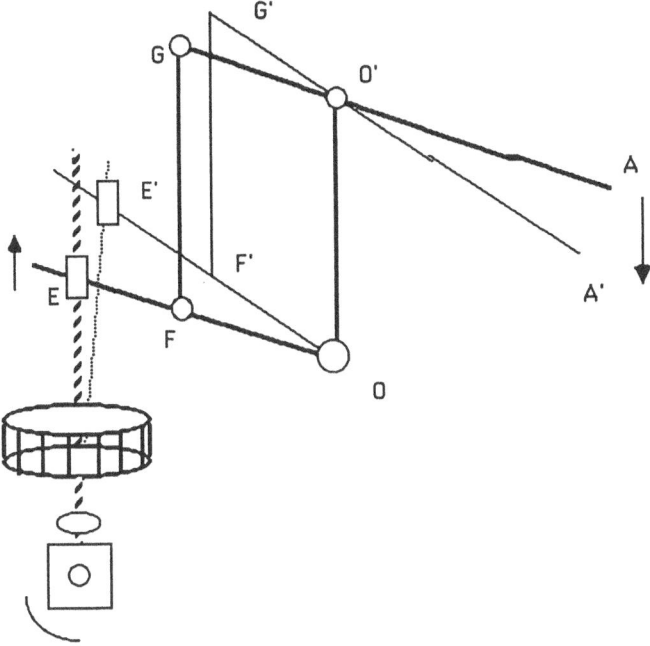

Figure 6b. Working parallelograms

2. Conclusion on Geometrical Activities in these Microworlds

The work we have done shows how interesting learning environments we can build with theses ideas. The very impressive motivation shown by the pupils when working with theses tools is an encouragment to do more.

We can find here ways to approach the problem of motivation to learn geometry. We are sure that even if the goal is to leave the real world, we have to start from it. This is specially true for beginners who require learning slopes from concrete world to abstract worlds. We reach like this, practical ways to solve the problems of sense: we offer possibilities to build links from formal systems to interpretations (only them carry sense). This is necessary to convince beginners about the need for formal systems. We so reach environments where it is possible to justify the need for modelling, explain the attitudes in modelling. It can come a clear separation between modelling by physicists (who extract laws from observations, measures) and modelling by mathematicians (who formally describe properties, structures, and use theorems to acquire certainty about the world). The real need for proving can appear here.

The different possibilities offered by modelling and functions available with models can also be examined. For example, we can predict moves, phenomena, events, compute and verify results, examine phenomena at their limits,...

This fits well with historical aspects around geometry: we are not so far from geometry to represent Egyptian fields or the trajectories of cannon balls for artillerymen.

The problem of teacher training still remains with this kind of open tools. This is very important to allow acquisition of the pre-geometrical culture carried out by theses tools. More the problem of tutoring the student in such spaces is not solved. Are ITS adequate tools to do this? The question remains open.

Knowledge Based Tutors for Geometry

Intelligent Tutoring Systems for geometry (Geometry Tutor [Anderson 85], MENTONIEZH [Py 89]) are designed to train students to be able to draw figures and build proofs, and to make them learn theorems and their use. In our opinion, the use of such systems suppose the learner to have a good understanding of the concepts we have just discussed about models, geometrical situations,... ITS are generally built on knowledge based architectures. This allows the system to solve the problems submitted to or by the tutor during the student sessions and to reach good possibilities for explanations.

1. Some Aspects of Intelligent Tutors in Mathematics

Different characteristics appear as fundamantal to justify the use of knowledge based architecture to write so-called ITS:

1) An expert system can be used as a *problem solver* in the domain and this makes it able to solve exercises submitted to the learner or by the learner. The expression 'competent tutor' is sometimes used to underline this possibility. This allows pedagogical strategies in which the learner proposes challenges to the system: exercises to be solved taking account of given constraints. Writting the solver imposes to make explicit useful knowledge to reach a given objective and in itself is very important for teaching.

This approach is very interesting but remains possible only for advanced students, able to set problems. For geometry, different kinds of solvers can be built: solver to do constructions of figures according to constraints [Buthion 75], provers able to prove geometrical properties in a 'human-like way'.

2) An expert-system can '*explain*' the solution it gives to a problem. Nowadays, most of expert-systems based tutors are limited to an analysis of knowledge of a given domain, an implementation of it under a production rules form or a semantic net of structured objects. Even if given in a nice aspect, explanations are more or less restricted to a trace of the items (rules) applied during the solving process by an inference engine. We need more to reach interesting pedagogical possibilities. Approaches using graphics (Geometry Tutor [Anderson 85]) to interact with the learner seem to be very effective but the type of explanation carried out in this way must be cleared.

The meaning of the word 'explanations' must be explained. A first typology allows separation between *trace, comments, explanations,* and *justifications:*

- **Trace** is restricted to the output of used items to solve a problem (trace of applied rules). We cannot speak of explanations as there is no 'understanding', no 'interpretation', no formulation or reformulation by the system of what is said.
- **Comments/Display** consists in revealing essential phases, the presentation of the logical articulations of inference sequences. This can be delivered from a syntactical analysis of the trace of applied rules. For example [Safar 85] has used the concept of 'pertinent facts' to prune the inference tree.
- **Explanations** have to be reserved for *comments* in which references to the contents, the semantics of the rules are done.
- **Justifications** imposes explicitation to the user of *how* and *why* were made the choices of knowledge used to solve a problem, explicitation of how, why a given strategy has been chosen, why not an other one,...

To obtain interesting tutors, we need to reach the *justification* level. This can only be done if we use Knowledge on knowledge. The formalisation of this upper level knowledge leads to a so called meta-knowledge. The system must explicitly know what it knows, how to use its knowledge, how to perform the control and choices while building a solution. It must be given under a declarative form (meta-rules). The ambition of *justification* imposes to implement effectively declarative meta-knowledge. We have done this kind of work in our KEPLER shell; so tracing, commenting, explaining,... metarules uses is a way to work towards very powerful tutors. It deals with transmission of strategies to solve problems in the field of the concerned matter. It supposes that it's possible to explain how to use the transmitted knowledge to make it efficient.

3) The tutorial component itself can also have an expert system architecture. This implies we have the possibility to write pedagogical rules to fill up the corresponding knowledge base. This implies also we have the possibility to write the meta-knowledge relevant when choices dealing with the control of the session have to be made.

4) Student modelling: the need for student modelling is now well recognised to allow flexible/personalised interactions. A lot of work have been done now and even if a first typology for uses of student models is now available [Self 88a], the problem remains very difficult. Possible attitudes include trying pragmatic approaches [Self 88b] or theoretical ones [Self 89]. A clear separation between *use* of a student model and *construction* of a student model must be made. It's also clear that modelling the students Laurel and Hardy is different from modelling classes of students, designing prototypes for subsets of learners. The easier is perhaps to only use prototypes of learners. How this can be done for geometry?

We can define classes of learners like thoses who are unable to understand the text of the problem, thoses who are unable to draw a figure representing the exact givens for the problem, those who are unable to separate between the givens and the goals,... According to these classes, tools like MENTONIEZH [Py 89], CABRI-Géomètre [Baulac 88] or Geometry Tutor [Anderson 85] can be helpful. But, how to diagnose the student and find the right class for Laurel and Hardy?

More, we not only have to take account of the learner, but we have to take account of the *context* of the learning process.

Which tools are available to help the learner? (rule and compass or CABRI-Géomètre?...) How does the learner works? Is he alone? With a teacher? With a fellow (co-learner working on the same problems)?

2. KEPLER

We have written the KEPLER [Vivet 88c] shell, useful to write expert systems, specially intelligent tutoring systems. Knowledge is given under a procedural form (procedure which can be called from rules) and a rule form. There are three categories of rules in KEPLER:

- **rewriting rules** to encode sure knowledge in a given domain;
- **production rules** to encode know-how and heuristics. They have the semantics:

    < pb pattern > $\Rightarrow$ < pattern of a plan to solve pb >

    They encode pre-determined *plans* (the major part of expertise) describing scenarios to be tried to solve the problem represented by <pb pattern>.
- **meta-rules** useful to choose the right plan to solve a given problem.

3. AMALIA

We are writting the AMALIA system, devoted to teach algebraic manipulations, to train students solving problems like integration of numerical functions.

The retained architecture is based on the cooperation of two KEPLER based expert-systems: a KEPLER based problem solver (CAMELIA [Vivet 84]) and a KEPLER based pedagogical module used in the tutorial component to conduct the session with appropriate pedagogical rules.

We give in [Vivet 88c, 88f] the general organisation and the architecture of AMALIA, a system under development to train people on algebraic manipulations problems.

The work focuses mainly on two fundamental aspects:

1) the problem of explanations at the *justification* level

2) the problem of pedagogical expertise. We re-use here rules coding pedagogical knowledge under a form of plans. The KEPLER shell is used to implement the tutoring part of AMALIA.

4. Hierarchy of Knowledge

Several kinds of knowledge items must be managed in such systems. A work we have done is to separate clearly between several types of knowledge and meta-knowledge useful to implement a knowledge based tutor.

The result is a description of a hierarchy between them. Basic relations between theses types are now elicited. For example, pedagogical rules can work on meta-rules useful to control the solver before firing it on a given exercise to be solved. This allows adaptation of the solution being built for the student according to an available model for the learner. So, if a given student has not yet seen a given concept, a pedagogical rule can make available for the solver a meta-rule excluding temporarily rules based on this concept. This prevents the solver from presenting an incomprehensible solution to the student. The current typology of knowledge is presented in [Vivet 87b] showing the technical possibilities we thus reach to take account of the student model. This allows more flexibility and adaptation to the learner. Examples of rules and meta-rules we have written for the AMALIA system are given in [Vivet 88c].

5. Conclusions on our Approach to ITS

1) AMALIA is still under development, but we have now real and interesting possibilities to deliver explanations with the solver. The tutorial part is not yet developed enough to imagine the kind of balance, atmosphere, interactions, possibilities of changes in styles of interactions,... we will reach with such a system. But the tools are ready and the needed possibility of coding tutorial strategies in a declarative way is now clear.

2) But the system is not used for geometry; what can be done? The feeling we have is that geometry is very specific because of the role of the figure. It seems important to give the learner tools to help him (her) understanding the text of the problem, to give him/her tools (draw sheets), to make him/her trying something, conjecturing. Finally it's important to give him/her guidance during the construction of a proof, help to remind him/her useful theorems,... So the question is now: 'How to merge ideas from MENTONIEZH, CABRI and Geometry Tutor?' But we have not forgotten that this activity supposes that the learner is motivated for geometry, for the proving attitude.

Figures 7a,b,c on the following page give some possible organisations to begin a discussion around theses ideas. The problem of producing a pertinent realisation is still largely open.

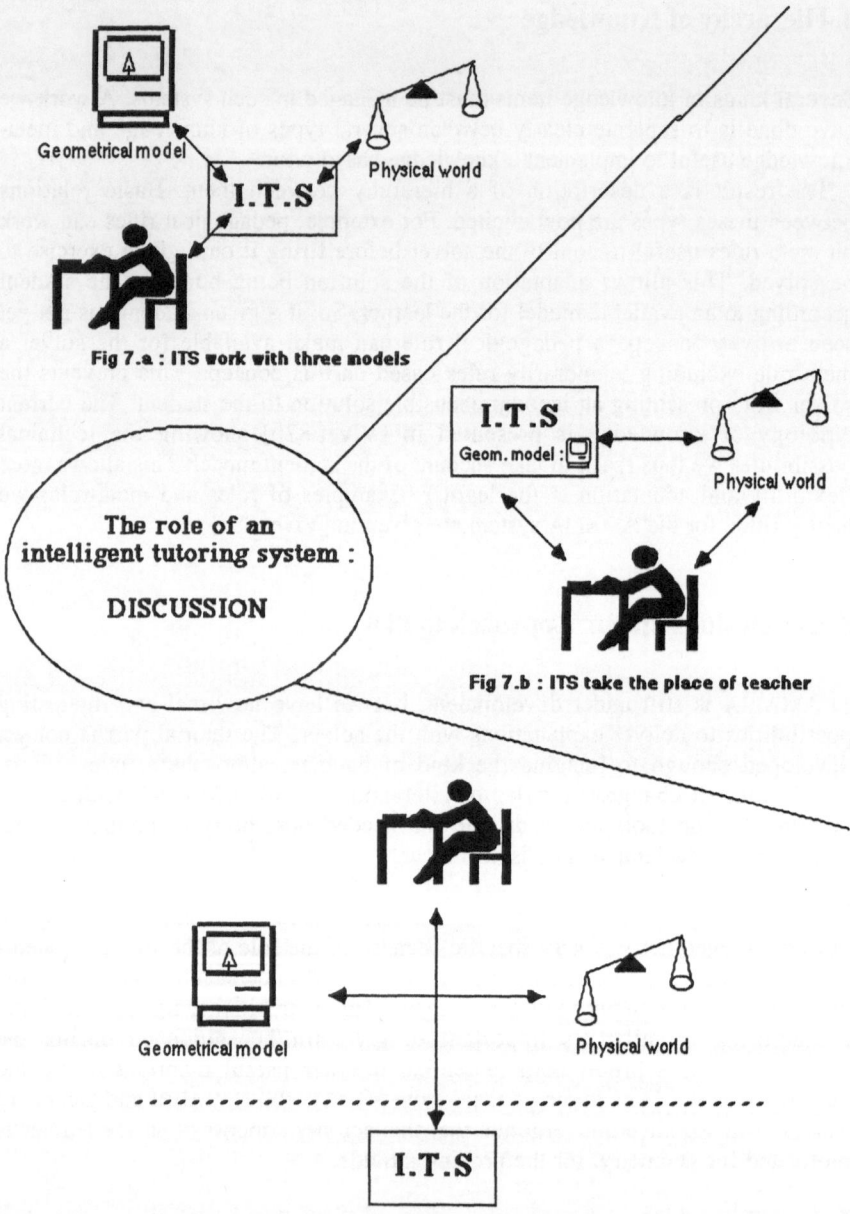

Fig 7.a : ITS work with three models

Fig 7.b : ITS take the place of teacher

Figure 7c. ITS as a background for the construction of knowledge

Main Conclusion

We have worked on two completely different approaches of the use of computers in education. We need deeper work in the two complementary ways. But it is possible now to consider the role of each approach in education. As claim by S. Ohlsson, education of pupils imposes acquisition of 'concepts' and 'procedures'.

Our feeling is that ITS fit better to teach procedures and micro-worlds are more relevant to learn concepts.

- ITS imposes student modelling: OK for procedures.
- ITS are important to be more operational in teachers training. The feedback from the work done on ITS for teacher training is not yet strong enough.
- ITS underline the need for a better definition of the role of the teacher with ITS. As the time for a replacement of the teacher by an ITS is yet very far, teachers will have to cooperate with ITS. So the study of their role is important and can be helpful in the design of ITS.

And robot based micro-worlds are:

- important for basic concepts acquisition,
- important to keep contact with the reality and the approach of basic problems like modelling,
- important to train student with inductive approaches, to make them imaginative.

Some questions remain:

What do we have to model the student knowledge in geometry? Is it possible? If the answer is yes, which knowledge must be coded? Known theorems? Buggy theorems?

Do we have to know about the representation of the concept of theorem for the learner? Does my learner perceive a theorem as an operational tool? Does he/she know the importance of the verification of all the hypothesis to apply a given theorem?

References

Part 1: Learning environments

Papert S. (1980) Mindstorms: children, computers and powerful ideas, Basic Books, New York.

Baulac Y., Bellemain F., Laborde J.M. (1988), CABRI-géomètre, un logiciel d'aide à l'enseignement de la Géométrie, logiciel et manuel d'utilisation, Cedic-Nathan.

Vivet M. (1986), Pilotage de Micro-robots sous LOGO, un outil pour sensibiliser les personnels de l'industrie à la robotique - 5e symposium canadien sur la technologie pédagogique. Ottawa, mai 1986. Paru dans le livre "A l'école des robots", robothèque du CESTA, p. 195-210.

Vivet M. (1986), Driving Micro-robots under LOGO: a way to approach geometry. Proceedings of the second international conference for LOGO and maths education. Institute of Education. University of London, July 1986, pp. 216-225.

Vivet M. (1989), Which goal, which pedagogical attitudes with micro-robots in a class room? NATO Advanced Research Workshop on Student Development of Physics Concepts: The Role of Educational Technology, Pavia, Italy, October 1989.

Part 2: ITS

Anderson J.R., C.F.Boyle, Yost G., (1985), The geometry tutor, proceedingd of the 9th International Joint conference on artificial intelligence, IJCAI 9, August 1985.

Baron M., Vivet M. (1988). systèmes experts et tuteurs intelligents, Congrès AFCET - Reconnaissances des formes et intelligence artificielle, Antibes, November 1988.

Buthion M. (1975), Un programme qui résout formellement des problèmes de constructions géométriques, thèse 3ème cycle, Université de Paris VI, 1975.

Buthion M. (1979), Un programme qui résout formellement des problèmes de constructions géométriques, RAIRO Informatique, vol. 13, n°1.

Chouraqui E., Inghiltera C., Veronis J. (1988), Conception d'une base de connaissances orientée objets pour l'EAO de la géométrie. Actes de l'Université d'été "Intelligence artificielle et enseignement des mathématiques", Toulouse, 1988.

Clancey W.J. (1983) . The epistemology of a rule based Expert system - a framework for explanations, Artificial Intelligence 20, 215-251.

Genesereth (1982). The role of plans in intelligent teaching systems. In: D.Sleeman-J.S.Brown eds., Intelligent tutoring systems, Academic Press, p.134-155.

Guin D., Rousselot J.F. (1987), Recherche en vue de la réalisation d'un programme d'aide à la démonstration en géométrie, actes COGNITIVA 87, Paris.

Kassel G. (1986). Expliquer, c'est raisonner sur le raisonnement: le système CQFE, VI ièmes journées internationales Systèmes experts et Applications, Avignon 1986, vol 2, p. 973-990.

Kassel G., Parchemal Y., Pitrat J., Safar B.(1986) Pourquoi utiliser des métaconnaissances? . Journées nationales sur l'intelligence artificielle, PRC-IA, GRECO-CNRS, Aix-les-bains, 20/21 nov 1986, Cepadues éditions, p. 79-88.

Mandl H., Lesgold A.M. (eds) (1987) Learning issues for intelligent tutoring systems, Springer-Verlag, New York.

Nicaud J.F., Vivet M. (1988). les tuteurs intelligents: réalisations et tendances de recherche, synthèse sur les tuteurs intelligents, revue Techniques et Science Informatiques -TSI, vol 7, n°1, janv. 1988.

Nicolas P., Allen R., Trilling L. (1987), Specification logique de figures pour l'aide à l''enseignement de la géomètrie, COGNITIVA 87, Paris, p. 26-32.

Nicolas P. (1989), Construction et vérification de figures géométriques dans le système MENTONIEZH, Thèse de l'Université de Rennes 1.

O'Shea T., Bornat R., Du Boulay B., Eisenstadt M., Page I. (1984). Tools for creating intelligent computer tutors. In: Elithor, Banerjii (eds), Human and artificial intelligence, North-Holland.

Ohlsson S. (1986). Some principles of intelligent tutoring. In Instructional Science 14, .293-326.

Pitrat J. (1986). connaissances et métaconnaissances. In intelligence des mécanismes et mécanismes de l'intelligence. Fondation DIDEROT. FAYARD (ed. J.L.Lemoigne), p. 75-113.

Py D. (1989), MENTONIEZH: an ITS in geometry, 4th international conference on AI & education, Amsterdam, May, pp. 202-209.

Ross P. (1988). Plan recognition for intelligent tutoring systems, Workshop IFIP-TC3, Frascati, mai 1987, in Artificial intelligence tools in education, North-Holland.

Safar B. (1985). Explications dans les systèmes experts, Journées systèmes experts, Avignon 1985.

Self J. (1986). Artificial intelligence, Its potential in education and training, 5ième symposium canadien sur la technologie pédagogique, Ottawa, May 1986, pp. 69-77.

Self J. (1988a). Student models: what use are they?, Workshop IFIP-TC3, Frascati, May 1987, in Artificial intelligence tools in education, North-Holland.

Self J. (1988b).Bypassing the intractable problem of student modelling, ITS 88, Montreal, juin 88, pp. 18-24.

Self J. (1989).The case of formalising student models (and ITS generally), 4th International conference "AI and Education", Amsterdam, May 1989.

Sleeman D. and Brown J.S. (1982). intelligent tutoring systems, Academic Press, London.

Vivet M. (1987a). systèmes experts pour enseigner: méta-connaissances et explications, congrès CESTA: MARI-COGNITIVA 87, Paris, 18/22 mai 1987.

Vivet M. (1987b). Hierarchy of knowledges in an intelligent tutoring system; how to take account of the student, note interne CRIC, Presented during the European seminar on Intelligent tutoring systems, Tübingen, October 1987.

Vivet M. (1988a). Reasoned explanations need reasoning on reasoning and reasoning on the student, Workshop IFIP-TC3, Frascati, mai 1987, in Artificial intelligence tools in education, North-Holland (1988).

Vivet M., Futtersack M., Labat M. (1988b). Métaconnaissances dans les tuteurs intelligents, international conference Intelligents Tutoring Systems, Montréal, juin 1988.

Vivet M. (1988c), Knowledge Based Tutors: Towards the design of a shell..., in International Journal for Educational Research and Instruction, 12(8), 839-850, Special issue edited by H. Mandl.

Vivet M. (1988d), la prise en compte du contexte avec les tuteurs intelligents, conférence invitée, Premier congrès européen "intelligence artificielle et formation", APPLICA 88, Lille, octobre 1988.

Vivet M., Carrière E., Delozanne E. (1988e), presentation of different aspects of AMALIA: a knowledge based system for mathematics, European summer university on intelligent tutoring systems, Le Mans, October/November 1988.

Vivet M. (1988f), Research in advanced educational technology: two methods. In: E. Scanlon, T. O'Shea (eds.), New directions in educational technology. NATO ASI Series F, Vol. 96. Springer-Verlag, Berlin.

Wenger E. (1987), Artificial Intelligence and tutoring systems Computational and cognitive Approaches to the Communication of knowledge, Morgan Kaufmann.

Winans R.T., Whitaker E.T., Bonnel R.D. (1988), Theories of learning in computer-aided Instruction, The fifth International Conference on Technology and Education, Edinburgh, March 1988, p. 86-89.

Yazdani (1986) . Intelligent tutoring systems survey, AI Review 1, 43-52.

Complex Factors of Generalization Within a Computerized Microworld: The Case of Geometry

Michal Yerushalmy

Laboratory for Research and Development of Computers' Uses for Learning
The University of Haifa, School of Education, Haifa, Israel
E-mail: REDC410@uvm.haifa.ac.il

The rationale for teaching geometry as part of the standard high school curriculum is twofold: 1) to teach students about the measurement, properties, and relationships of points, lines, angles, surfaces, and solids; and 2) to teach students deductive reasoning by exposing them to classical Euclidian geometry, the archetype deductive system. Most geometry courses come up short on both counts.

The centerpiece of most geometry instruction is neither the 'stuff' of geometry nor deductive thinking, but the two column geometric proof which in many respects seems to be beyond the grasp of many students. Students cope by memorizing theorems and proofs and with no understanding and appreciation of either geometry or deductive reasoning and proof.

We argue that geometry instruction would be more effective if, rather than teaching definitions and theorems as given and concentrating on proof, it were to give students an opportunity to experiment with geometric shapes and elements, to move from the particular to the general, and to make conjectures before grappling with formal proofs. This approach to geometry is absent from the secondary geometry curriculum. With the infusion of inquiry skills and a tool such as the GEOMETRIC SUPPOSER (Schwartz & Yerushalmy 1985/88) into geometry learning and teaching such an approach is feasible.

In this approach which we call 'guided inquiry', the content of the curriculum is the same as a standard geometry course but the goals are different. Rather than focusing only on deductive reasoning and proof, the guided inquiry approach calls for students to integrate inductive reasoning with deductive reasoning and empirical work with conceptual understanding while solving problems and devising proofs.

While designing the mode of work within such a microworld it is assumed that students would be motivated to explore, gather and analyze many examples and by that doing create a basis for self convincing and acceptance of a collection of incidents as mathematical phenomena. In other words, we are expecting that generalization would occur among other inductive and deductive reasoning. According to various heuristics (Polya 1954, Schoenfeld 1986, Brown & Walter 1983) geometry problem solvers, whether dealing with a construction or a proof, have to plan, look for similar problem, check impact of various variables, etc. It is therefore expected that generalization within the suggested a microworld would improve problem solving skills.

Our studies present evidence that the ability to generalize while working within a microworld is a function of various factors and it will be discussed below.

Definitions

Two types of generalizations:

Induction is a well known process to reach generalizations by the examination of instances or examples.

An instance or a set of instances is examined, certain properties are identified.

The given example is then taken as a member of a larger set and its properties are input to the larger set. Such generalization from multiple examples is developed on a similarity based approach (Chi & Bassok 1989).

Another type of generalization is a process acted out on a statement (either a conjecture or a proven statement). Such generalization requires the relaxing of conditions within the statement and therefore called in the literature condition-simplifying generalization (Holland et al. 1986). Studies investigating the conditions for the evolution of such generalization from a single idea tie the ability to generalize with the tendency and the ability to construct an explanation to the single given statement (Chi & Bassok).

Both types of generalizations are methods to create a type of conjecture and neither is a process for deriving definite knowledge. In both, the end result is a statement whose truth is unknown for the time it is made.

Generalizations and Conjectures

People reach conclusions by various processes. One of them is conjecturing, whereby one offers a statement that one thinks may be true, though at the time one doesn't know for sure. Such a statement is a conjecture. By calling a statement a conjecture, or saying that it resulted from a person's conjecturing, imply no particular process of creation. A conjecture can result from explanation, belief, experience, deductive proof, or generalization. Generalizations are a particular kind of conjecture, created by reasoning from the specific to the general.

Geometric conjectures have three key parts: the relationship described in the conjecture, the set of objects for which the relationship holds, and the quantifier which determines the members of the set of objects for which the relationship holds. Sometimes one or more of these parts is not explicitly stated, but is understood.

Since a geometrical statement usually included more than the verbal description of the problem (e.g., a diagram or numerical information) there are situations in which our distinction, which is related to the source of the generalization, is difficult to apply. First, when a numerical aspect of a statement is generalized, the word 'induction' seems to be an appropriate, or natural, description of the process. In this view, the initial statement is one example, and any statement which substitutes a different numerical value is another example. The general statement is then reached by examining the specific cases. Second, in situations when a diagram and a particular statement are given, it is difficult to know whether the person addressing the problem is working from the example or from the statement. To draw conclusions in such a case, one must infer what is taking place in the mind of the conjecturer. While our distinction is difficult to apply in these situations, we feel that it is valuable to distinguish among the different ways generalizations are created.

Do Learners Encounter any Difficulties in Forming Generalizations?

Nickerson, Perkins, and Smith (1985) argue that the ability to test theories seems to be more prevalent than the ability to construct them. More specific obstacles are rooted in the tendency of people to limit their hypothesis space to oversimplified situations, instead of considering all possible hypotheses that can be formed from a sample. The well known bias of inductive reasoning, both in creating samples and in forming conjectures is the confirmation paradox; Holland et al. (1986) found that a belief in a specific hypothesis usually leads to sampling towards the confirmation of the hypothesis. No special effort is made to find disconfirming evidence. When a non- instance occurs spontaneously, it is ignored in order to allow confirmation. This is a repeated finding that presents people tendency to create an hypothesis that relates any two factors X and Y by observing only the cases in which both are valid, while ignoring the other cases.

The studies which dealt with generalization in mathematics mostly concentrated on the reasons for over-generalization. Matz (1982) found that errors occurring in high school algebra problem solving resulted from reasonable attempts to generalize previous knowledge into a new situation.

Studies of visualization in geometry (Rissland 1977, Hoz 1981, Yerushalmy & Chazan 1990) found that for most geometry students, diagrams are intended as models, but since people usually only sample upon availability and mostly from memory, the models or the standard diagrams serve as the sample for generalization and bias its results. Hershkowitz et al. (1987) findings suggest that the beliefs and the concept images that students carry dominate the geometrical evidence and reality. Consequently, students do not look for negative cases or 'non-examples' of a concept and easily generalize perception into wrong concept or definition.

Chi and Bassok (1989) suggest that plausible generalizations occur only when the subject is involved in self-explanation of the example. Thus, examples provided by any microworld, even carefully planned, are a necessary condition for formation of theories but not a sufficient one. Other factors such as previous concept images and beliefs and the general ability and knowledge of the content are involved as well.

Evaluating Generalizations

A Conjecture/Generalization Test

The conjecture's test was one of the data sources for observing types of and measuring level of generalizations in geometry in addition to classroom observations, students written works, argument test and recordings of the teachers' as well as students perspectives about the work (Yerushalmy, Chazan & Gordon 1987).

A pretest and a post-test were designed to assess students' ability to make conjectures or general statements (see Yerushalmy 1986 for complete details). The tests present students with problems that are composed of a statement and diagram(s) that illustrate the statement. Problems on the tests are posed as data formulations and abstract formulations. In data formulations, the statements contain data and are designed to provide insight into students' ability to generalize based on given instances. In the abstract formulations, the statements are designed to provide insights into students' ability to derive plausible categories for generalizations.

The tests ask students to 'write significant connected statements' in response to the problem. Items for the pretest consisted of items assumed to be known from previous learning of geometry. For each problem, a list of attributes had been compiled using the analysis of similar problems in plane geometry by Brown and Walter (1983). Each list was divided to three parts, using the Structure-Mapping theory developed by Gentner (1983) and the definition of Spontaneous Analogy by Clement (1983). The three parts are:
  (i) geometric attributes,
  (ii) numerical attributes, and
  (iii) key (fixed) features.
The test was scored for the level of generalization, the type of attribute observed as a subject for generalization in each problem, the plausibility of the answer, its correctness and originality.

The scores that identified the major measures were:

Level of generalization, which answers the question: How general is the statement?
  No generalization (0): Specialization or no generalization.
  Unconnected (1–2): Addresses more general cases than in the given idea, but leaves the information unconnected. For example, if it mentions 4 parts in a given shape instead of the 2 parts given, and then mentions 6 parts as a different idea.
  Process (3): Generalizes a process but does not identify a general conjecture. For example, the organization of geometric information into the sequence 2,4,6,8.
  Phenomenon (4): Conjecture that describes a general phenomenon, such as even numbers, in any quadrilateral, etc.

Type (of variable changed in the original statement) scores four options for changes in attributes of a geometry problem that promoted generalization:

No change (0): Repeats the same given attribute.

Object (1): Replaces an object with a more general one. For example, any triangle instead of a right triangle.

Relation (2): Replaces a geometric relation. For example, an inequality between measurements of geometric objects converted into an equality.

Numerical (3): Changes numerical information into variables.

Fixed features (4): Revises key features.

The first two changes are strongly related to the 'idea generalization' while the third presents the 'generalization from data'.

Plausibility, which answers the question: How important was the contribution of the new idea to the original one?

Not related (0): Does not relate to the idea in the statements at all.

Trivial (1): Relates trivially (a rectangle has four sides).

Valid (2–3): Valid as a generalization to the statement but does not add originality to the statement (the sum of exterior angles in any rectangle is 360°).

Original (4): The new idea is not included in the curriculum.

All scores for plausibility were related to the curriculum covered by the participants before taking each of the tests.

Qualitative Evaluation

Beyond a quantitative analysis using the above scoring mechanism we do individual tests to identify possible types of processes aimed towards a certain level of generalization. The following are visual descriptions of processes of generalizations carried out by high school students[1] .

One type of process is presented in Figure 1. This type of generalization is characterized by the large amount of examples explored before reaching a higher level of generalization. Such students progress in small steps towards generalization and often end up with producing many examples on the same level and even ended up with a specialization instead of a generalization.

Another type of generalization (such as the one presented in Figure 2) occurs when students start with a general statement (often over-generalized) and then observe less general cases: examples or ideas which are more concretely connected to the given statement.

We also identified students who looked for only a single example at each of the levels of generalization.

---

[1] The test asked for generalizations of given statements. The statements discussed here are:

1) In a rectangle, the interior and exterior angles are equal.

2) A line passing through the center of the square and parallel to one of its sides divides the square to equal areas.

3) On a peg board (a geoboard), a right triangle can be drawn by using 6 pegs.

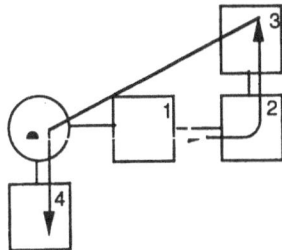

Problem #1

1. In a square the inrerior angles are 90°.
2. An interior and exterior angle that share a side form a liner pair.
3. Two sides of a polygon if extended would form two pair of vertical angles.
4. In a square interior and exterior are congruent.

Problem #2

1. Can create a square with 4 pegs.
2. A rectangle with 6 pegs.
3. An isosceles using 4 pegs.
4. If you attach a rubber band which extends the hypotenuse from the midpoint you create perpendicular bisector and divide the triangle into 2 equal parts.

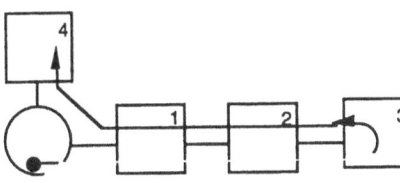

Figure 1. A typical process of generalization drawn from a few examples at the same level

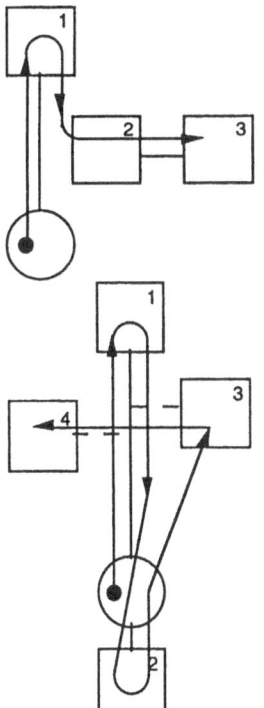

Problem #1

1. Any line passing trough the center of the square will divise into 2 equal parts.

2. If a line passes trough the center is parallel to the sides it is a line of symmetry.

Problem #2

1. An infinite # of right triangles can be made on the board.

2. The right triangle with 6 is isosceles.

3. Using the triangle numbers [1, 3, 6, 10, 25, 21, 28,...] all right triangles can be isosceles.

Figure 2. Immediate over-generalization followed by a few examples

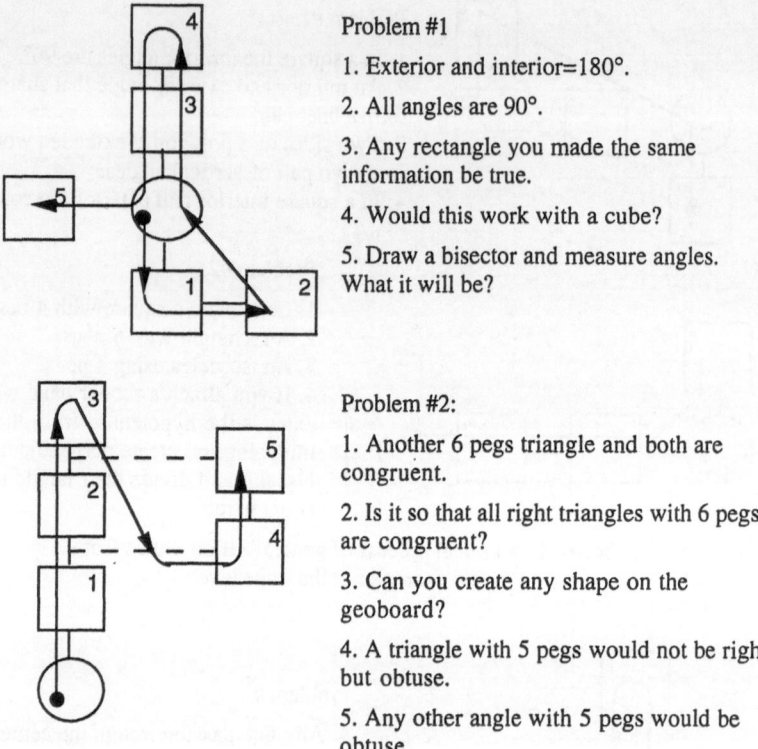

Problem #1

1. Exterior and interior=180°.

2. All angles are 90°.

3. Any rectangle you made the same information be true.

4. Would this work with a cube?

5. Draw a bisector and measure angles. What it will be?

Problem #2:

1. Another 6 pegs triangle and both are congruent.

2. Is it so that all right triangles with 6 pegs are congruent?

3. Can you create any shape on the geoboard?

4. A triangle with 5 pegs would not be right but obtuse.

5. Any other angle with 5 pegs would be obtuse.

Figure 3. Generalization reached by an example for each level

Through this analysis we tried to look for a model of generalization and seem to find that this model should be a function of ability level, task type and personal style and as so is difficult to be defined.

Individual interviews should be taken in order to examine those findings.

Features of the Microworld to Support Generalization

The SUPPOSER was designed to focus and direct users' attention by providing certain options and not others, by making some things easy to do while making others difficult or impossible.

For the sake of the current analysis specific software options must be isolated from the whole, while in reality the impact of the software is the combined impact of all of the options.

Choosing initial shape: built in categories for sampling

One important aspect of the design of the SUPPOSER is the way shapes (triangles or any other polygon) are classified. The user must specify the initial shape. If the user chooses one of the predetermined categories, a random shape of that kind (random sized and where possible random relationship between sides and angles) appears on the screen in a random orientation. This characteristic of the tool can challenge students misconceptions about shapes and about constructions related to a certain class. If the user does not choose one of the predetermined categories, he can control the creation of the initial shapes using YOUR OWN option. This characteristic of the SUPPOSER allows the user the freedom to test conjectures by creating extreme cases that are candidates for counter examples.

Construction tools: availability of visual information

After choosing the initial shape, the user can produce a geometrical construction. The construction tool allows the user to create any Euclidian construction quickly and simply and provide primitives, like 'median', that are less primitive than pure compass and straightedge construction. By reducing the overhead for creating accurate geometrical constructions, this aspect of the software makes it possible to teach and learn based on large quantities of visual information.

Repeating on a process: observing generality

In the SUPPOSER's menu-driven environment the user specifies the desired construction by choosing a menu item and using correct, formal geometric language to describe, without ambiguity, where the construction is to be made. Thus, the SUPPOSER can capture, as a procedure, all of the constructions carried out on an initial shape. The REPEAT option allows the user to try this procedure on a new or previous initial shape, thus reducing the construction burden even further. This option frees the user from the single diagram and allows him to track characteristics of a construction that are invariant from shape to shape.

Quantitative information

Within the SUPPOSER numbers are used in two ways: in measurements and in constructions that request specification of length, size of angle, etc. Within each of the two uses the numbers have two meanings: numbers which present on screen measurements and quantifiers which present a geometric property by ratios. Here are a few examples to clarify this distinction.

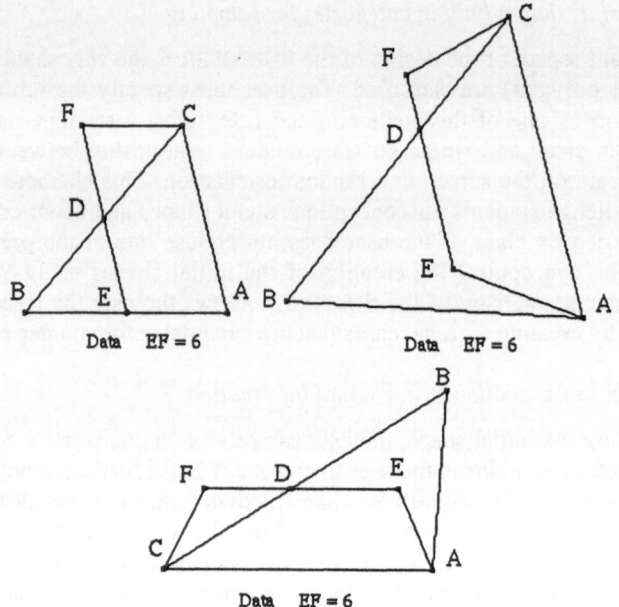

Figure 4. Construction while using an absolute length (EF = 6 units)

Figure 4 presents a construction of a parallel segment in the length of 6 screen units, thus forming a parallelogram EACF. When repeating the same construction on a different triangle (Figure 4a) or even when rescaling the diagram (Figure 4b) the construction looks geometrically different.

Thus, focusing on the local and non relevant property of concrete length, the shape of the construction changes its properties. In Figure 5 the length of the parallel line was specified by related measure to the side AB (AB/EF = 1). Doing so maintains the geometric properties while varying the initial shape.

Finally, it must be emphasized that the SUPPOSER does not stand alone; it is part of an approach to teaching geometry. The students work with the software is a part of the course, not the whole.

Therefore, as important or even more important than the software itself is how its use is integrated into the course and how teachers make use of the capabilities the software provides.

Our study of the implications of inquiry activities to the generalizations carried out with the SUPPOSER (Yerushalmy, Chazan & Gordon 1987) suggested that students explored problems as they were written. They did no more than the instructions suggested and they were unwilling to expand or change a problem unless generalization was a part of the activity requirements.

For example, testing the following two problems, the first version yielded fewer generalizations than the second.

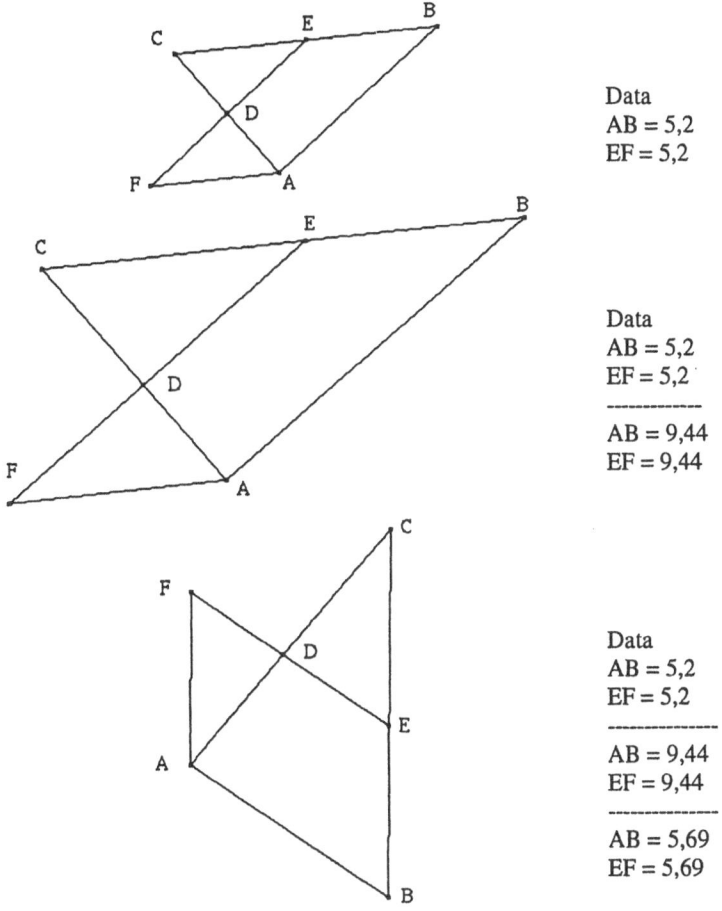

Figure 5. Construction while using a relative length (EF = AB)

Version 1

Try to split a triangle into triangular sections that have the same area. First try to get four sections with equal area. When you have a solution make sure it works in all kinds of triangles. Record a drawing of your solution and explain what kind of lines you added to the triangle.

Version 2

One segment can cut or divide any triangle into triangular sections. For example, an angle bisector divides any triangle into two sections, and three medians divide any triangle into six sections. Give the types and number of line segments that divide any triangle into 2,3,4,5 sections all having the same area.

Our observations made it clear that there was a relationship between the formulation of the problems and their success in motivating generalizations.

What Are the Skills that Need to be Developed?

Without getting into pedagogical and ability differences, we will now discuss the various considerations students have to take when sampling and analyzing towards general conjectures while operating with the SUPPOSER (Yerushalmy 1993).

Sampling of Data

The use of the SUPPOSER to allow easy sampling did not automatically free students from the need to create samples confirming their previous or naive generalizations nor allowed them to ignore the information that did not confirm their hypothesis.

Consideration of non-stereotypic instances

People often fail to generalize appropriately while sampling stereotypic instances. To determine whether the SUPPOSER might have an effect on students' ability to overcome this obstacle we compared SUPPOSER scores (Yerushalmy 1991) and results obtained by non-SUPPOSER classes (both groups were 8th graders of a comparable ability). Among other items of basic geometric concepts (Hershkowitz 1989) students were asked to identify right angle triangles among 9 diagrams of triangles. Three of the nine were right triangles and two of the three were in a non-standard position. Significant differences were found on the overall performance of this problem between the SUPPOSER students and the comparison group (T = 2.68, p < 0.01). Similar results were obtained at other tasks such as drawing altitudes in non-vertical positions and while identifying polygons.

While we attribute those results to the capability provided by the SUPPOSER to view shapes of a well defined class in a random appearance, it should be noticed that the natural tendency of novice inquirers was to seek control over the information whenever the 'random' members of the class did not meet their concept images. They even chose to reconstruct the whole construction step by step one on a new triangle without using the REPEAT option because they refused to believe to what they see. Only then, finally convinced that the data provided by the 'uncontrolled' REPEAT option was correct. Thus, the availability of the tool does not guarantee the appreciation of data when the information does not support the original expectations.

Consideration of the size of the sample

How much geometric data is suitable for sampling in learning geometry? This is an especially hard question to relate to since in most mathematics classes, the amount of work required of the students in order to complete a certain task is clearly specified and since the traditional learning of geometry does not make use of large quantities of information (one diagram is often considered sufficient).

In our pedagogical approach, we defined the amount of information needed as a function of the plausibility of the data and therefore the amount collected became more of a personal decision than a standard.

We followed students' efforts to deal with it (Yerushalmy 1990) and found that students were concerned about the size of their sample and often produced an irrelevant quantity of instances.

Analyzing the Data

The manipulation of numerical data

Being able to compare numbers is essential for looking for patterns. Comparing numbers by observing differences and ratios was therefore intensively used while manipulating samples. While the technique itself was well known to high school students, we collected evidence that misconceptions or insufficient knowledge about the structure of numbers caused false numerical manipulations, which affected the quality of induction. For example, one student could not find an expected pattern because she unintentionally computed the ratios of two perimeters using two different methods of representation. Students confident in their algebraic ability used larger samples of numerical information more often than others. The inability to formulate a general algebraic formula for patterns caused difficulty in stating the generalizations. The conjectures were either false generalizations or applicable to a very limited number of cases. Additional directions within the task formulation helped such students to reach a categorization of their samples.

The manipulation of visual information

While there are standard techniques to compare and order numbers, they are not attainable within the visual domain.

The ability to overcome visual obstacles using the SUPPOSER is described in detail in another work (Yerushalmy & Chazan 1990). Here, we briefly present the main findings related to visual manipulations. SUPPOSER users in our studies learned to reorganize the visual field of a given situation. By so doing they were able to see various data within the same picture. Thus, the developed visual skill improved the quality and the quantity of the generalizations. However, individual differences were found between what could be called 'visualizers' and 'non-visualizers' in deductive reasoning and generalizations.

In other previous works we discussed in detail the connection between the need to generalize and the understanding of deductive proof (Chazan 1989).

Observations providing repeated results about the tendency of SUPPOSER students to provide convincing arguments (Yerushalmy 1986, Schwartz 1989) suggest that there is a mutual effect between the confrontation with false generalizations and the development of the need for proof.

Generalizations from Ideas: A Teaching Intervention

A preliminary condition for generalizing from an idea is the ability to examine the idea's components and selectively relax them, thus open avenues for generalization. The following is a description of an experiment to teach a strategy leading to generalizations. We taught one of two comparable SUPPOSER groups (9th graders) to produce generalizations and to analyze geometric ideas through posing questions about each of the statement's attributes. The goal was to help students develop the ability to perceive a geometric idea as a versatile entity that could be changed or identified as a case of a class. The teacher in the experimental group used a strategy of problem posing rather than problem solving. She chose to do it in a way suggested by Brown and Walter (1983) which they call "what if not" strategy. Using this strategy one generalizes by relaxing conditions within a problem by asking "what if this condition (restriction) does not exist?" .

Figure 6 illustrates schematically the progression of geometric themes discussed and worked on in this class using the SUPPOSER, following the "what if not" strategy. When asked to conjecture about a different geometrical subject while working independently, these students used the strategy modeled by their teacher, and formulated many interesting questions and conjectures. At the end of the learning period a significant difference was obtained between the two groups in their ability to generalize (for more details see Yerushalmy & Maman 1988).

This strategy is not the only one. Based on this structure one may teach other units which will emphasize other general reasoning skills.

Summary and Discussion: Modeling Generalisers

We described here our work concerning generalization in geometry using the GEOMETRIC SUPPOSER. The work describes two approaches to generalization: induction or similarity based generalization and generalization of an idea or explanation based generalization. We identified difficulties and misconceptions raised while using both types of generalization and observed three major steps leading towards generalizations: finding potential phenomena for generalization from an idea, sampling major cases and analyzing samples.

Using the SUPPOSER and the curricular material developed to support the inquiry activities we observed and evaluated students performance. The generalization test helped to observe and compare levels and categories of generalizations. Classroom and individual observations helped us to identify major skills

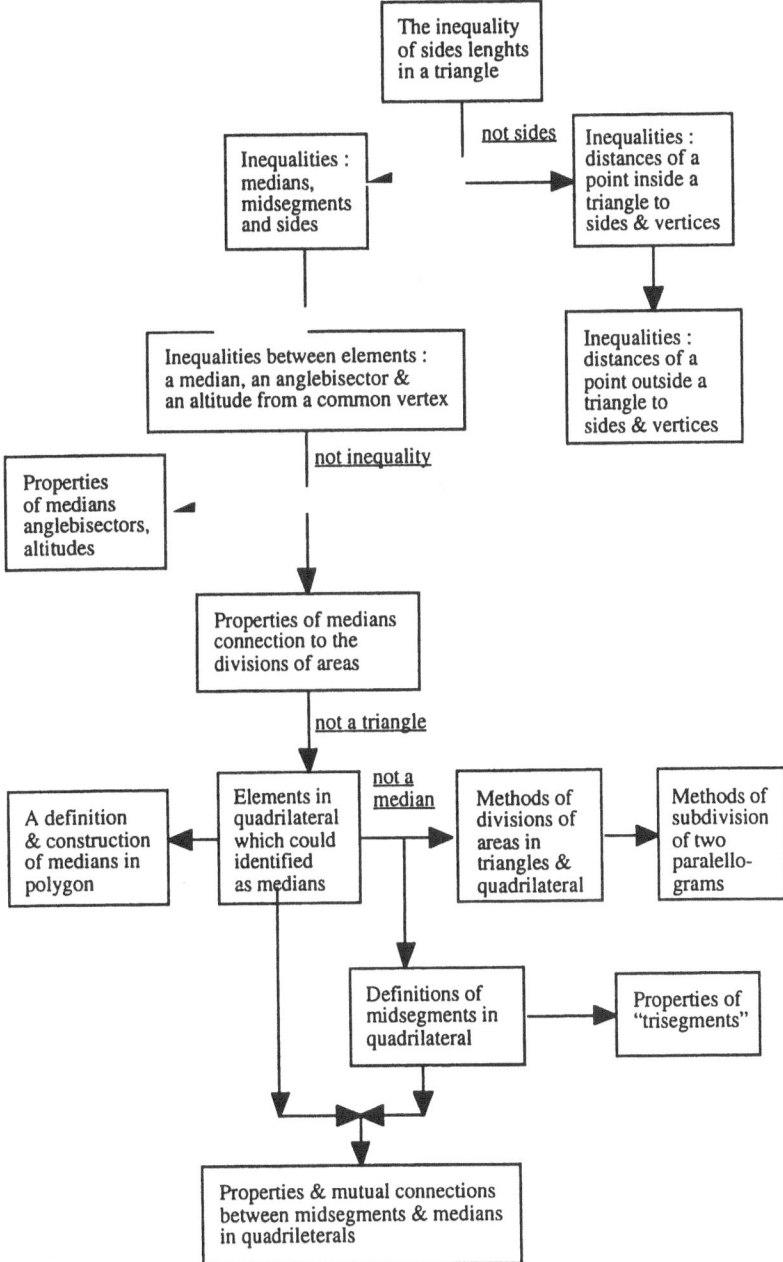

Figure 6. A flowchart of a generalization unit using the "what is not" strategy

involved in the process of generalization such as visual analysis, algebraic and deductive skills. As a result special units were developed to help students and teachers challenge obstacles.

Does all this work bring us closer to suggesting a students' model for generalization in geometry?

Since any answer should rely on additional evidence I will conclude, based on personal view of the nature of learning geometry and of student-teacher-machine relations, by pointing at the difficulties in creating such a model as a result of our work.

(i) Our experience suggests that the nature of the task has an enormous impact on the expected outcomes from the student. Dreyfus (this volume) suggests that the content domain has an impact on the ability to create a student model. I would sharpen this claim: using our evidence, a different phrasing of the same content and the same task had an impact on the form of generalizing.

(ii) We were not able to clearly identify the type of representation that had been the source for action in each case nor to exactly determine the type of generalization used: it was either a result of looking at many diagrams, performing induction on numerical data, or relaxing conditions that were stated verbally. We explained above the reasons for our interpretations but we did not try to isolate each action.

(iii) We found various paths towards both high and low level generalization.

We might suggest that factors other than ability of the student and the type of the task had an impact on this performance. A quote from Polya (1962) here might be persuasive: "in inductive reasoning one use shrewd reasons, personal feeling which cannot be measured" (p. 23).

We believe that this work which reflects on one major aspect within six years of observing and supporting the use of the SUPPOSER in the schools, would encourage further investigation of the possibility to enhance the learning within inquiry environment in geometry.

References

Brown S.I., Walter, M.I. (1983) The art of problem posing. The Franklin Institute Press. Philadelphia, PA.

Chazan, D. (1989) Ways of knowing: high school student's conceptions of mathematical proof. Unpub. Doctoral Dissertation, Harvard Graduate School of Education, Cambridge, MA.

Chazan, D. (1988b) Proof and measurement: an unexpected misconception. In: Proceedings of the PME XII, Hungary.

Chi, M.T.H., Bassok, M. (1989) Learning from examples via self-explanations. In L.B. Resnick (ed.) Knowing, learning, and instruction. Erlbaum Associates, Hillsdale, NJ.

Clement, J. (1983) Observed methods for generating analogies in scientific problem solving. Paper presented at AERA annual meeting, Canada.

Dreyfus, T., Schwarz, B. (1995) Cognitive interpretation of microworld operations. (In this volume).

Gentner , D., Gentner, D.R. (1983) Flowing waters of teeming crowds: mental models of electricity. In D. Gentner, A.L. Stevens (eds.) Mental models. Erlbaum Associates, Hillsdale, NJ.

Hershkowitz, R. (1987) The acquisition of concepts and misconceptions in basic geometry – or when "A little learning is a dangerous thing" in J.D. Novak (eds). Proceeding of the 2nd international Seminar Misconceptions and Educational Strategies in Science and Mathematics. Cornell Univ. Vol. III.

Holland J.H., Holyoak K.J., Nisbett R.E., Thagard P.R. (1986) Induction: processes of inference, learning and discovery. MIT Press, Cambridge, MA.

Hoz, R. (1981) The effects of rigidity on school geometry learning. Educational Studies in Mathematics 12, 171-190

Matz, M. (1982) Towards a process model for high school algebra errors. In: Sleeman and Brown (eds.) Intelligent Tutoring Systems. Academic Press.

Nickerson R.S., Perkins D.N., Smith E.E. (1985). The teaching of thinking. Erlbaum Associates, Hillsdale, NJ.

Polya, G. (1962). Mathematical discovery: on understanding, learning, and teaching problem solving. John Wiley and Sons, New York.

Polya, G. (1954), How to solve it. Princeton University Press, Princeton, NJ.

Rissland, E. (1977). Epistemology, representation, understanding, and interactive exploration of mathematical theories. Unpub. Doctoral Dissertation. Massachusetts Institute of Technology, Cambridge, MA.

Schoenfeld, A. (1986) On having geometric knowledge. In J. Hiebert (ed.), Conceptual and procedural knowledge: the case of mathematics. .Erlbaum Associates, Hillsdale, NJ.

Schwartz, J.L. (1989) Intellectual mirrors: a step in the direction of making schools knowledge-making places. Harvard Educational Review, 59(1), 3-13.

Schwartz, J.L., Yerushalmy, M. (1985-8) The GEOMETRIC SUPPOSER (Computer Software), Pleasantville, NY: Sunburst Communications.

Yerushalmy, M. (1986). Induction and generalization: An experiment in teaching and learning high school geometry. Unpub.doctoral thesis, Harvard Graduate School of Education.

Yerushalmy, M. Chazan D. (1990). Overcoming visual obstacles with the aid of the SUPPOSER. Educational Studies in Mathematics, 21, 199-219.

Yerushalmy, M. Chazan, D., Gordon, M. (1987). Guided inquiry and technology: a yearlong study of children and teachers using the GEOMETRIC SUPPOSER. Educational Technology Center Technical Report TR 88-6, Harvard Graduate School of Education.

Yerushalmy M. Maman H. (1988). Using the SUPPOSER for whole group explorations. Laboratory of Computers for Learning. Technical report #9 (in Hebrew). The University of Haifa, Israel.

Yerushalmy, M. (1991). Enhancing acquisition of basic geometrical concepts with the use of the SUPPOSER. Journal of Educational Computing Reseach, 7, 407-420.

Yerushalmy, M. (1993). Generalizations in geometry. In: Schwartz, J.L., Yerushalmy, M., Wilson (eds.) (1993). The GEOMETRIC SUPPOSER: What is it a case of? Erlbaum Associates, Hillsdale, NJ.

Contributors and Participants

Canada

Robert LA BARRE
Département Informatique, Université du Québec à Trois-Rivières
G9A 5H7 Trois-Rivières PQ
E-mail: Robert_Labarre@UQTR.UQuebec.CA

Marie-France THIBAULT
Département Mathématique, Université du Québec à Trois-Rivières
G9A 5H7 Trois-Rivières PQ
E-mail: Marie-France_Thibault@UQTR.UQuebec.CA

France

Nicolas BALACHEFF
DidaTech, IMAG-LSD2
BP 53, F-38041 Grenoble cedex 9
E-mail: Nicolas.Balacheff@imag.fr

Monique BARON
Laforia, 40 rue Bezout, F-75014 Paris
E-mail: baron@laforia.ibp.fr

Yves BAULAC
at time: IMAG-LSD2, BP 39, F-38041 Grenoble cedex 9
now: Diademe, 5 rue Tour de l'Eau, F-38400 Saint Martin d'Hères

Franck BELLEMAIN
DidaTech, IMAG-LSD2, BP 39, F-38041 Grenoble cedex 9
E-mail: Franck.Bellemain@imag.fr

Eugène CHOURAQUI
DIAM - IUSPIM, av. Escadrille Normandie Niemen
F-13397 Marseille cedex 20
E-mail: DIAM_EC@vmesa11.u-3mrs.fr

Bernard CORNU
at time: MAFPEN - RECTORAT
now: Dir. IUFM Grenoble, rue Marcellin Berthellot, F-38100 Grenoble

Georges FAFIOTTE
IMAG-GETA, BP 53, 38041 Grenoble cedex 9
E-mail: Georges.Fafiotte@imag.fr

Italo GIORGUTTI
Institut Mathématique de Rennes, IRMAR Campus de Beaulieu
Université de Rennes, F-35042 Rennes cedex
E-mail: giorgiut@univ-rennes1.fr

Michel GIRY
IRPEACS, BP 167, F-69131 Ecully cedex

Régis GRAS
Institut Mathématique de Rennes, . Campus de Beaulieu
Université de Rennes, F-35004 Rennes cedex
E-mail: Gras@univ-rennes1.fr

Dominique GUIN
at time: IREM Université Louis Pasteur
10 rue du Général Zimmer, F-67084 Strasbourg
now: ERES Département de Mathématiques, UM2
Place E. Bataillon, F-34095 Montpellier cedex 5
E-mail: guin@math.univ-montp2.fr

Carlo INGHILTERRA
at-time: GRTC CNRS, 31 chemin J. Aiguier, F-13402 Marseille cedex 9
now: DIAM - IUSPIM, avenue Escadrille Normandie Niemen
F-13397 Marseille cedex 20
E-mail: DIAM_CI@vmesa11.u-3mrs.fr

Colette LABORDE
DidaTech, LSD2-IMAG, BP 53, F-38041 Grenoble cedex 9
E-mail: Colette.Laborde@imag.fr

Jean-Marie LABORDE
LSD2-IMAG, BP 53, F-38041 Grenoble cedex 9
E-mail: Jean-marie.Laborde@imag.fr

Gérard LEJEUNE
CEPHAG ENSIEG, BP 46, F-38041 - Grenoble cedex
E-mail: glejeune@cephag.observ-gr.fr

Charles PAYAN
IMAG-LSD2, BP 53, F-38041 Grenoble cedex 9
E-mail: Charles.Payan@imag.fr

Laurent TRILLING
IMAG-LGI, BP 53, F-38041 Grenoble cedex
E-mail: Laurent.Trilling@imag.fr

Martial VIVET
Laboratoire informatique, Université du Maine
(Route de Laval), BP 535, F-72017 Le Mans cedex
E-mail: martial@lium.univ-lemans.fr

Germany

Gerhard HOLLAND
Institut für Didaktik der Mathematik
Karl Glöckner Str. 21C, D-35394 Giessen

Marion KONTÖP
Institut für Didaktik der Mathematik
Karl Glöckner Str. 21C, D-35394 Giessen

Heinz SCHUMANN
Pädagogische Hochschule Weingarten, Kirchplatz 2
D-88250 Weingarten

Rudolf STRÄSSER
IDM Bielefeld, Universität Bielefeld, Postfach 10031, D-33501 Bielefeld
E-mail: rstraess@hrz.uni-bielefeld.de

Israel

Tommy DREYFUS
Center for Technological Education, P.O. Box 305, Holon 58102
E-mail: ntdryfus@weizmann.ac.il

Baruch SCHWARZ
School of Education, Hebrew University, Mount Scopus, Jerusalem, Israel
E-mail: msschwar@mscc.huji.ac.il

Michal YERUSHALMY
Laboratory for Research and Development of Computer's Uses for Learning
The University of Haifa, School of Education, Haifa
E-mail: REDC410@uvm.haifa.ac.il

Italy

Maurizio FALCONE
Dipartimento di Matematica, Università di Roma "La Sapienza"
Piazzale A. Moro 2, I-00185 Roma
E-mail: falcone@axcasp.caspur.it

United Kingdom

Roshni DEVI
Centre for Information Technology in Education
The Open University, Walton Hall, Milton Keynes MK7 6AA

Sara HENNESSY
Centre for Information Technology in Education
The Open University, Walton Hall, Milton Keynes MK7 6AA

Celia HOYLES
Department of Mathematics, Statistics and Computing
Institute of Education, 20 Bedford Way London WC1H OAL
E-mail: choyles@ioe.ac.uk

Tim O'SHEA
Centre for Information Technology in Education
The Open University, Walton Hall, Milton Keynes MK7 6AA

Ronnie SINGER
Centre for Information Technology in Education
The Open University, Walton Hall, Milton Keynes MK7 6AA

Maria YANNISSI
Centre for Information Technology in Education
The Open University, Walton Hall, Milton Keynes MK7 6AA

USA

Richard ALLEN
St Olaf College
E-mail: allen@acc.stolaf.edu

Daniel CHAZAN
Education Research Center, MIT, Cambridge MA 02139
E-mail: dchazan@msu.edu

Anthony E. KELLY
Rutgers University, New Brunswick, NJ 08903

Eugene KLOTZ
Department of Mathematics (The Geometry Project)
Swarthmore College, Swarthmore, PA 19081
E-mail: klotz@forum.swarthmore.edu

Richard LESH
at time: ETS, Rosedale Road, Princeton NJ 08541
now: University of Massachussetts at Dartmouth, North Dartmouth, MA 02747
E-mail: rlesh@nsf.gov

Nicholas JACKIW
Key Curriculum Press, Berkeley, CA 94702
E-mail: njackiw@keypress.com

Judah L. SCHWARTZ
Educational Technology Center, Graduate School of Education
Harvard University, Cambridge, MA 02138
E-mail: judah@HUGSE1.bitnet

Baruch SCHWARZ
Educational Technology Center, Graduate School of Education
Harvard University, Cambridge, MA 02138
E-mail: msschwar@mscc.huji.ac.il

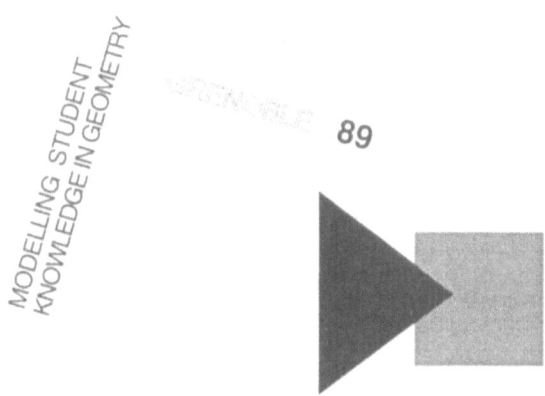

The NATO ASI Series F Subseries on
ADVANCED EDUCATIONAL TECHNOLOGY

NATO ASI Series F

NATO ASI Series F

Including Special Programmes on Sensory Systems for Robotic Control (ROB) and on Advanced Educational Technology (AET)

NATO ASI Series F

NATO ASI Series F

NATO ASI Series F

NATO ASI Series F

Including Special Programmes on Sensory Systems for Robotic Control (ROB) and on Advanced Educational Technology (AET)

NATO ASI Series F

Springer-Verlag
and the Environment

We at Springer-Verlag firmly believe that an international science publisher has a special obligation to the environment, and our corporate policies consistently reflect this conviction.

We also expect our business partners – paper mills, printers, packaging manufacturers, etc. – to commit themselves to using environmentally friendly materials and production processes.

The paper in this book is made from low- or no-chlorine pulp and is acid free, in conformance with international standards for paper permanency.